Frontiers in African Business Research

Series Editor
Almas Heshmati, Jönköping International Business School, Jönköping, Sweden

This book series publishes monographs and edited volumes devoted to studies on entrepreneurship, innovation, as well as business development and management-related issues in Africa. Volumes cover in-depth analyses of individual countries, regions, cases, and comparative studies. They include both a specific and a general focus on the latest advances of the various aspects of entrepreneurship, innovation, business development, management and the policies that set the business environment. It provides a platform for researchers globally to carry out rigorous analyses, to promote, share, and discuss issues, findings and perspectives in various areas of business development, management, finance, human resources, technology, and the implementation of policies and strategies of the African continent. Frontiers in African Business Research allows for a deeper appreciation of the various issues around African business development with high quality and peer reviewed contributions. Volumes published in the series are important reading for academicians, consultants, business professionals, entrepreneurs, managers, as well as policy makers, interested in the private sector development of the African continent.

Kennedy Modugu • John Otalor
Editors

Sustainability and Digital Transformation in Africa

Springer

Editors
Kennedy Modugu
Middlesex University
Flic en Flac, Mauritius

John Otalor
University of Calabar
Calabar, Nigeria

ISSN 2367-1033 ISSN 2367-1041 (electronic)
Frontiers in African Business Research
ISBN 978-981-95-9499-3 ISBN 978-981-95-9500-6 (eBook)
https://doi.org/10.1007/978-981-95-9500-6

This Springer imprint is published by the registered company Springer Nature Singapore Pte Ltd.
The registered company address is: 152 Beach Road, #21-01/04 Gateway East, Singapore 189721, Singapore

This book is dedicated to Africa's present and future stewards of progress: the scholars who interrogate power with evidence, the innovators who design solutions under constraint, the educators who cultivate critical minds, and the policymakers who carry the weight of intergenerational responsibility. It honors those who insist that technology must serve humanity, that growth must respect ecological limits, and that development must be inclusive, ethical, and just.

We dedicate this volume to the millions across the African continent whose lives are shaped by decisions made at the intersection of digital innovation and sustainability, often without their voices being heard. May this work contribute, in some small measure, to amplifying those voices and to shaping digital futures that expand opportunity without eroding dignity or the natural environment.

Finally, this book is for the next generation of African thinkers, builders, and leaders. In their hands lies the task of transforming digital promise into sustainable reality, and of ensuring that Africa's technological future is not merely efficient, but humane, equitable, and enduring.

Preface

The accelerating convergence of digital transformation and sustainability represents one of the most consequential dynamics shaping contemporary development discourse. Across the globe, digital technologies are increasingly framed as instruments for addressing long-standing economic, social, and environmental challenges. In Africa, however, this convergence assumes a particular urgency and complexity. The continent's demographic profile, ecological vulnerability, institutional diversity, and uneven digital infrastructure mean that digital transformation is neither a neutral process nor an automatic pathway to sustainable development. It is a deeply political, ethical, and strategic endeavor whose outcomes will depend on how technologies are designed, governed, and embedded within local contexts.

This volume, *Sustainability and Digital Transformation in Africa*, was conceived in response to the growing gap between global narratives of technological optimism and the lived realities of African societies. While artificial intelligence, digital platforms, data analytics, and automation are often presented as universal solutions, their implications in Africa are mediated by structural constraints, historical legacies, and pressing sustainability challenges. At the same time, Africa is not merely a recipient of digital innovation. It is a site of experimentation, adaptation, and indigenous problem-solving, where digital tools are increasingly mobilized to address climate risk, educational access, entrepreneurial inclusion, and communicative legitimacy.

The chapters selected for this book reflect this dual reality. Together, they offer a rigorous and interdisciplinary examination of how digital technologies are reshaping sustainability pathways across the continent. Rather than celebrating technology in abstract terms, the contributors interrogate its concrete applications, limitations, and unintended consequences. They ask not only whether digital transformation can support sustainability, but under what conditions it does so, for whom, and at what cost.

The volume brings together scholars working across diverse disciplinary traditions, including information systems, accounting and finance, entrepreneurship, education, linguistics, and ethics. This diversity is intentional. Sustainability challenges are inherently systemic, and digital transformation cuts across institutional and sectoral boundaries. The chapters therefore engage with multiple domains, from AI-driven business processes and entrepreneurial ecosystems to sustainability education and digital communication practices. In doing so, they highlight both complementarities and tensions between technological efficiency, social inclusion, environmental stewardship, and ethical governance.

A defining feature of this book is its emphasis on critical engagement. Several chapters explicitly address the risks associated with rapid digitalization, including ethical blind spots, regulatory gaps, cultural misalignment, and the reproduction of inequality. Others examine the conditions under which digital tools can be harnessed to promote green growth, inclusive entrepreneurship, and future-ready skills. Collectively, the contributions resist simplistic narratives of technological determinism and instead foreground agency, institutional design, and normative responsibility.

The editorial process underpinning this volume was intentionally rigorous. Each chapter underwent multiple rounds of peer review and revision, with particular attention to conceptual clarity, methodological transparency, and policy relevance. Only chapters that demonstrated both scholarly depth and alignment with the book's overarching purpose were selected for inclusion. The result is a cohesive collection of work that speaks to international audiences while remaining grounded in African realities.

This book is intended for a wide readership. Scholars will find theoretically informed analyses that contribute to ongoing academic debates on digital transformation and sustainability. Policymakers and development practitioners will encounter evidence-based insights that can inform regulatory design, educational reform, and innovation strategy. Educators and students will find material that supports critical engagement with the promises and pitfalls of digital technologies in development contexts.

Ultimately, this volume seeks to contribute to a more nuanced understanding of Africa's digital future. It affirms that digital transformation, when guided by ethical principles, contextual awareness, and long-term sustainability goals, can play a meaningful role in addressing some of the continent's most pressing challenges. At the same time, it underscores that technology alone is insufficient. Sustainable digital transformation requires deliberate choices, inclusive governance, and sustained investment in human and institutional capacity.

It is our hope that this book will serve not only as a scholarly reference but also as a catalyst for more thoughtful, responsible, and context-sensitive approaches to digital innovation in Africa.

Mauritius, Mauritius Kennedy Modugu
Calabar, Nigeria John Otalor

Acknowledgments

The completion of this volume has been made possible through the collective efforts of many individuals and institutions, and the editors wish to express their sincere gratitude to all who contributed to its intellectual and practical realization.

First and foremost, we extend our deepest appreciation to the chapter authors whose scholarship forms the core of this book. Their willingness to engage critically with complex questions at the intersection of digital transformation and sustainability, and to subject their work to rigorous review and revision, reflects a high level of academic integrity and commitment. The depth, diversity, and originality of their contributions have been central to shaping a volume that speaks both to African realities and to global scholarly debates.

We are equally indebted to the anonymous reviewers who devoted considerable time and expertise to the evaluation of submitted chapters. Their detailed, thoughtful, and often challenging feedback played a decisive role in strengthening the conceptual clarity, methodological rigor, and coherence of the final manuscript. The quality of this book is, in no small measure, a testament to their professionalism and intellectual generosity.

Our gratitude also extends to the editorial and production teams for their guidance and support throughout the development process. Their responsiveness, technical expertise, and attention to detail ensured that the manuscript progressed smoothly from conception to completion, while meeting the demanding standards of an international academic publication.

The editors would like to acknowledge the academic institutions and research communities that provided the intellectual environments in which this work was developed. These spaces of dialogue, mentorship, and scholarly exchange were instrumental in sustaining the long and often demanding process of editing an interdisciplinary volume of this nature.

Finally, we recognize the personal support networks of colleagues, mentors, friends, and families whose encouragement and understanding made it possible to bring this project to fruition. Academic work of this scale is never an individual endeavor, and it is sustained as much by human support as by intellectual effort.

To all who contributed, directly or indirectly, we offer our sincere thanks.

Abbreviations

AI	Artificial Intelligence
API	Application Programming Interface
AU	African Union
CBDC	Central Bank Digital Currency
CSRD	Corporate Sustainability Reporting Directive
DI	Digital Infrastructure
ESG	Environmental, Social, and Governance
FinTech	Financial Technology
GDP	Gross Domestic Product
GHG	Greenhouse Gas
ICT	Information and Communication Technology
IFRS	International Financial Reporting Standards
IoT	Internet of Things
ISSB	International Sustainability Standards Board
KYC	Know Your Customer
ML	Machine Learning
NGO	Non-Governmental Organization
PRISMA	Preferred Reporting Items for Systematic Reviews and Meta-Analyses
R&D	Research and Development
RPA	Robotic Process Automation
SDGs	Sustainable Development Goals
SLR	Systematic Literature Review
SMEs	Small and Medium-Sized Enterprises
STI	Science, Technology, and Innovation
TAM	Technology Acceptance Model
TBL	Triple Bottom Line
UN	United Nations
UNCTAD	United Nations Conference on Trade and Development

UNECA	United Nations Economic Commission for Africa
UNESCO	United Nations Educational, Scientific and Cultural Organization
WEF	World Economic Forum

Contents

1 **Understanding Africa's Technological Ecosystem** 1
Kennedy Modugu and Tasneem Mustun

2 **From AI Workflow Automation to Acceleration: The Impact of Artificial Intelligence on Sustainable Business Growth in Africa** 31
Francis Okoye

3 **AI-Driven Entrepreneurship: Catalyzing Green and Inclusive Growth in Africa** 55
Dennis I. Edogun

4 **Ethics of Artificial Intelligence in Africa's Sustainability Agenda: Risks, Challenges, and Policy Considerations** 81
John Ikponwosa Otalor and Irfaan Oodally

5 **Sustainability Education in the Digital Age: Building a Future-Ready Workforce** 113
Shahira Abdalla and Eseroghene Udjo

6 **Digital Communication Practices and Sustainability Messaging in African Contexts: A Cultural Linguistics and Pragmatic Perspective** 135
Dare Eriel Ehigie and Jimoh Junior Braimoh

List of Figures

Fig. 2.1 Three-Stage Evolution of AI Workflow Automation and Sustainability 38
Fig. 3.1 The Integrated Conceptual Framework for Sustainable AI in Africa 73
Fig. 4.1 The United Nations Sustainable Development Goals 84
Fig. 4.2 A Framework for Linking the SDGs with AI Impact 85
Fig. 4.3 Ancient Myths About Artificial Intelligence 86
Fig. 4.4 Agenda 2063—"The Africa We Want": Technology and Innovation as Drivers of Inclusive Growth, Good Governance, Cultural Renaissance, and Peaceful and Secure Societies 94
Fig. 5.1 Conceptual Framework for Digital Sustainability Education in African Higher Education 118
Fig. 5.2 PRISMA Flow Diagram 124
Fig. 6.1 Integrated Analytical Model for Digital Sustainability Messaging in African Contexts 141

List of Tables

Table 5.1 PRISMA Search Strategy 122
Table 5.2 MMAT Quality Appraisal Summary by Study Type 125
Table 6.1 Comparative Overview of Digital Sustainability Messaging Practices in Nigeria, Kenya, and Ghana 156

Contributors

Shahira Abdalla Business Department, Higher Colleges of Technology, UAE, Email: sosama@hct.ac.ae

Jimoh Junior Braimoh University of Mississippi, USA, Email: jimoh2braimoh@gmail.com

Dennis I. Edogun Conestoga College, Canada, Email: denogun@yahoo.com

Dare Eriel Ehigie State University of New York at Buffalo, USA

Kennedy Modugu Middlesex Business School, Middlesex University, Mauritius Campus, Email: K.modugu@mdx.ac.uk

Tasneem Mustun Middlesex Business School, Middlesex University, Mauritius Campus

Francis Okoye Lagos Business School, Pan-Atlantic University, Email: fokoye@lbs.edu

Irfaan Oodally Middlesex Business School, Middlesex University

John Ikponwosa Otalor Department of Accounting, University of Calabar, Email: John.otalor@unical.edu.ng

Eseroghene Udjo Business Department, Higher Colleges of Technology, UAE

Chapter 1
Understanding Africa's Technological Ecosystem

Kennedy Modugu and Tasneem Mustun

Abstract Africa's technological evolution reflects a complex interplay of history, policy choices, and entrepreneurial agency, shaped by shifting economic, political, and social conditions. This chapter offers an integrated analysis of Africa's technological ecosystem through three interrelated lenses: historical trajectories, policy and institutional frameworks, and business-led innovation. It traces Africa's technological foundations from precolonial indigenous knowledge systems through colonial disruption, post-independence industrial ambitions, and the restructuring effects of neoliberal reforms. The chapter argues that these historical legacies continue to shape contemporary innovation systems and patterns of technological dependency. Adopting a policy perspective, the chapter examines the development of national and continental science, technology, and innovation frameworks, highlighting variations in institutional capacity, governance coherence, and implementation effectiveness across African countries. Comparative insights from selected countries illustrate how differing policy regimes and institutional arrangements influence innovation outcomes. The analysis further explores the rise of business-led innovation and the digital economy, focusing on the roles of entrepreneurship, venture capital, public–private partnerships, and emerging technologies such as fintech, artificial intelligence, blockchain, and renewable energy. The chapter also assesses the socioeconomic implications of Africa's technological transformation, including productivity growth, employment dynamics, inclusion, and inequality, while critically engaging with persistent structural constraints such as infrastructure deficits, financing gaps, and reliance on foreign technology. In conclusion, the chapter proposes strategic pathways for building a more inclusive, resilient, and self-sustaining technological ecosystem. By situating technological change within Africa's unique historical and institutional contexts, the chapter moves beyond simplistic narratives and highlights African agency in shaping the continent's technological futures.

K. Modugu · T. Mustun
Middlesex Business School, Middlesex University, Flic en Flac, Mauritius
e-mail: k.modugu@mdx.ac.uk; t.mustun@mdx.ac.mu

K. Modugu, J. Otalor (eds.), *Sustainability and Digital Transformation in Africa*, Frontiers in African Business Research,
https://doi.org/10.1007/978-981-95-9500-6_1

Keywords Technological ecosystem · innovation · pre-colonial Africa · infrastructure deficit

1 Introduction

Africa's technological evolution has been neither linear nor uniform; it is a story of resilience, adaptation, and creative response to shifting economic and social realities. From the earliest evidence of indigenous iron smelting in Nok culture to the continent's current prominence in mobile banking and renewable energy innovation, Africa's engagement with technology has reflected a dynamic interplay between necessity and ingenuity. Today, Africa is at a remarkable inflection point. While still grappling with structural challenges such as infrastructure deficits, skill gaps, and limited research funding, the continent has also become a frontier for technological experimentation. The rise of the digital economy, anchored on mobile connectivity, fintech, agritech, e-commerce, and renewable technologies, signals a shift from mere technology consumption to active innovation and creation (United Nations Economic Commission for Africa [UNECA], 2021).

Over the past two decades, the expansion of mobile telephony, particularly through platforms such as Flutterwave in Nigeria and M-Pesa in Kenya, has dramatically transformed access to finance and social inclusion. Internet penetration and smartphone adoption have opened new pathways for entrepreneurship, remote work, and e-governance. Similarly, the rise of innovation hubs in cities like Lagos, Nairobi, Kigali, Accra, and Cape Town illustrates how technology can drive job creation and knowledge economies. Yet, despite these encouraging developments, the African technological ecosystem remains marked by contradictions. While some regions pioneer in artificial intelligence or blockchain-based solutions, others still struggle with basic access to electricity or broadband. The coexistence of high-tech innovation and persistent underdevelopment defines Africa's technological reality and underscores the need for nuanced and historically informed analysis.

Understanding Africa's technological environment requires more than a snapshot of its current state; it calls for an exploration of the trajectories that shaped it. Historically, Africa's encounter with technology has been deeply intertwined with colonialism, postcolonial policy experiments, and global economic forces. Colonial regimes introduced technologies primarily for extraction and control, not for endogenous development. Post-independence governments sought industrialization through import substitution and state-led planning, but these efforts were often constrained by dependence on imported machinery, foreign expertise, and limited institutional capacity. The neoliberal reforms of the 1980s and 1990s further restructured African economies, introducing market-oriented policies that reshaped technological priorities toward privatization and foreign investment. These historical phases continue to influence the continent's present innovation systems and policy environments.

A policy lens is equally critical because technological progress is not self-generating; it is shaped by governance frameworks, institutional coordination, and strategic investment in research and development. The diversity of science, technology, and innovation (STI) policies across African countries reveals varying levels of state commitment and institutional maturity. For example, South Africa's well-established National System of Innovation (NSI) contrasts with weaker policy coherence in many other regions. By examining these policy regimes, one can identify what has worked, what has stagnated, and what reforms are necessary to build robust innovation ecosystems.

Similarly, an innovation lens captures the entrepreneurial dynamism that has defined Africa's 21st century transformation. Grassroots inventors, start-ups, and social enterprises have become engines of inclusive innovation, often addressing local challenges such as access to energy, healthcare, and markets with context-specific solutions. Understanding how innovation unfolds in these local contexts amid regulatory constraints and resource scarcity offers valuable lessons not only for Africa but for global theories of innovation and development.

Together, these three perspectives of historical, policy, and innovation allow for a comprehensive and interdisciplinary understanding of how Africa's technological identity has been constructed and how it continues to evolve in the face of global change.

The primary objective of this chapter, therefore, is to reveal the complexity of Africa's technological evolution by tracing its historical roots, assessing its institutional and policy underpinnings, and examining the rise of business-led innovation across sectors. It aims to move beyond the simplistic narratives of "Africa rising" or "Africa lagging behind" and instead situate technological change within the continent's unique historical, political, and socioeconomic contexts. By doing so, the chapter highlights the agency of African societies in shaping their technological futures while acknowledging the structural constraints they continue to face.

Following this introduction, Section 2 examines the historical paths of African technology, from indigenous knowledge systems to colonial legacies and post-colonial industrialization efforts. Section 3 assesses the policy regimes and institutional frameworks that have shaped technological progress, encompassing science and innovation policies, intellectual property rights, and regional cooperation. Section 4 highlights business innovation and the digital economy, demonstrating how African entrepreneurs and private sector players are redefining technological boundaries. Section 5 examines the socioeconomic impacts of these developments, including job creation, inclusion, and inequality. Section 6 discusses ongoing challenges and structural constraints, including insufficient infrastructure, funding gaps, and reliance on foreign technologies. The final section, Section 7, proposes pathways for the future, offering actionable strategies to build a more inclusive, self-sustaining, and resilient technological ecosystem across Africa. Section 8 concludes the chapter.

2 Historical Trajectories of Technology in Africa

2.1 Precolonial Indigenous Technologies

Long before European contact, Africa possessed sophisticated systems of knowledge and technological ingenuity that were deeply embedded in its ecological and cultural contexts. Precolonial African societies developed technologies that were not only functional but also sustainably designed to harmonize with local environments and community needs. Contrary to the colonial portrayal of a "technologically backward" continent, archaeological and anthropological evidence reveals that indigenous innovation was widespread, diverse, and adaptive across regions (Holl, 2009; Phillipson, 2005; Schmidt & Childs, 1995).

Agricultural systems were among the most advanced aspects of precolonial African technology. In regions such as the Sahel and the Nile Valley, Africans developed irrigation, terracing, and crop rotation systems that sustained large populations under varying climatic conditions. The terraced fields of Ethiopia's highlands, the flood-recession farming of Mali's Niger River delta, and the banana- and yam-based agroforestry systems in West and Central Africa exemplify deep ecological intelligence. These systems integrated soil conservation, water management, and biodiversity preservation centuries before the rise of modern sustainability discourse. Seeds were selectively bred for resilience, and intercropping was used to maintain fertility and minimize pest damage. These technological practices were born from generations of observation and adaptation.

Metallurgy was another area of significant innovation. African blacksmiths and smelters developed efficient and eco-friendly techniques for extracting and working iron. Archaeological discoveries at sites like Nok (Nigeria), Meroë (Sudan), and Tanzania's Ufipa region show that sub-Saharan Africans independently invented iron smelting as early as 1000 BCE, using furnaces that reached high temperatures without bellows. The resulting iron tools and weapons improved agriculture, trade, Yoruba Kingdoms in West Africa, Kongo Kingdom in Central Africa, Aksumite Empire in East Africa and strengthening the economic and political power of kingdoms such as Benin, Mali, and Great Zimbabwe in Southern Africa. Metallurgy also had a spiritual aspect, as blacksmiths were honored as guardians of transformative power, symbolizing the blend of material and metaphysical knowledge.

Architecture and construction technologies reflected environmental adaptation and aesthetic sophistication. The Pyramids of Giza, the stone structures of Great Zimbabwe, the coral-built towns of Swahili city-states along the East African coast, and the earthen mosques of Timbuktu and Djenne demonstrate mastery of engineering principles suited to local climates. These constructions used renewable materials such as mud, coral, stone, limestone and timber, while incorporating climate-sensitive features such as natural ventilation and thermal regulation. Architecture was also a form of social expression as compounds and cities were designed to embody communal life, hierarchy, and spirituality, illustrating a holistic approach to technology that merged utility with cultural meaning.

Equally important were Africa's knowledge traditions, including systems of medicine, astronomy, and navigation that reflected practical understanding of natural phenomena. Indigenous healers used extensive botanical knowledge to practice pharmacology, while astronomical observations helped guide agricultural calendars and navigation across the Sahara and along the coastlines. Oral traditions, apprenticeships, and craft guilds played a key role in passing this knowledge down through generations. These systems were not static; they evolved constantly, adapting to ecological changes and trade interactions. Essentially, precolonial Africa's technologies were deeply rooted innovations, connected to the environment, community, and spirituality, and serving both economic needs and cultural identity.

2.2 Colonial and Postcolonial Transitions

The advent of European colonialism in the late 19th century marked a profound rupture in Africa's technological trajectory. Colonial administrations systematically reoriented technology toward extraction, control, and dependency rather than endogenous development. Infrastructure such as railways, ports, and telegraph lines was built not to integrate African economies internally but to facilitate the export of raw materials to Europe and the import of manufactured goods. This extractive infrastructure model entrenched technological asymmetries that persist to this day.

During colonial rule, technology transfer was highly selective and hierarchical. Africans were excluded from scientific education and high-skill industrial training; instead, they were relegated to manual labor and low-level technical roles. Colonial industries prioritized mining, plantation agriculture, and primary processing, leaving indigenous industries and crafts to decline under competition from imported goods. Traditional metallurgy, textile production, and indigenous architecture were either discouraged or marginalized, labeled as "primitive" in colonial discourse. This technological subjugation ensured that innovation remained externally directed and locally dependent.

Missionary and colonial schools introduced Western curricula that devalued indigenous knowledge and replaced it with Eurocentric models of science and technology. Consequently, Africa's intellectual infrastructure became oriented toward replication rather than invention—a legacy that still influences postcolonial education systems. The colonial state's monopoly over research institutions and its control of technology through licensing and patent systems established a dependency pattern that outlived colonial rule.

Following independence in the 1950s and 1960s, African leaders recognized technology as a key to economic sovereignty and modernization. Early postcolonial states pursued industrialization through import-substitution strategies, seeking to replace imported goods with domestically manufactured ones. Governments established state-owned enterprises, research institutes, and technical universities to promote self-reliance. Countries such as Ghana under Nkrumah, Egypt under Nasser, and Tanzania under Nyerere launched ambitious industrial projects in steel, textiles, and machinery, often with support from the Soviet Union or China.

However, these initiatives faced formidable challenges. Many postcolonial economies lacked the capital, skills, and managerial capacity to sustain heavy industries. Moreover, dependence on imported machinery and spare parts perpetuated reliance on foreign technology. Political instability, weak institutions, and limited markets further undermined industrial performance. The state-led industrialization model, though visionary, became unsustainable amid fiscal crises and global economic shocks of the 1970s. Nevertheless, it represented an important phase in Africa's pursuit of technological autonomy, a time when the state actively sought to harness technology for nation-building and social transformation.

2.3 The Structural Adjustment Era

By the early 1980s, Africa entered a new era of economic restructuring driven by the Structural Adjustment Programs (SAPs) imposed by the International Monetary Fund (IMF) and World Bank. These policies aimed to liberalize markets, privatize state enterprises, and reduce public spending, with profound implications for technological development (International Monetary Fund, 2003). While intended to promote efficiency and attract foreign investment, SAPs instead hollowed out Africa's nascent industrial and research capacity. Public funding for universities, laboratories, and innovation agencies was drastically reduced, leading to brain drain, deteriorating infrastructure, and weakened institutional capacity.

The liberalization of markets opened African economies to global competition, but without corresponding protection for local industries. Domestic manufacturers and technology firms, which were already fragile, were quickly displaced by cheaper imports. Research institutions, deprived of funding, shifted their focus toward short-term consultancy work for donors, fragmenting the continuity of scientific inquiry. The state's withdrawal from industrial policy left a vacuum that private sectors, constrained by limited capital and expertise, struggled to fill. Consequently, Africa's technological dependency deepened, as multinational corporations dominated telecommunications, energy, and extractive sectors.

Yet, amid these constraints, new forms of informal and grassroots innovation emerged. The economic austerity of the SAP era forced individuals and communities to improvise solutions to everyday problems of repairing, repurposing, and reengineering materials creatively. In Nigeria's informal automotive sector, artisans developed expertise in fabricating spare parts; in Kenya and Ghana, small-scale mechanics built machinery from recycled components; and across West Africa, local innovators experimented with solar energy, water pumps, and radio technologies. This period gave rise to a culture of "frugal innovation"; innovation driven by necessity rather than abundance.

Informal innovators became the backbone of resilience in African technology ecosystems. While unrecognized by formal institutions, their adaptive problem-solving demonstrated the continent's latent capacity for endogenous innovation. This "bottom-up" creativity laid the groundwork for the entrepreneurial dynamism that would later characterize Africa's digital revolution in the 2000s. It also challenged the conventional Western model of innovation, showing that invention can emerge from scarcity, and not necessarily from surplus.

3 Policy Regimes and Institutional Frameworks

The evolution of Africa's technological environment cannot be understood without examining the policy regimes and institutional frameworks that shape how knowledge is produced, diffused, and applied. Technology does not develop in a vacuum; it is cultivated through the interaction of governance structures, legal systems, research institutions, and market forces. In Africa, these frameworks have evolved through complex historical phases: from the centralized, state-led industrial policies of the post-independence era to the market-oriented and partnership-driven models of the twenty-first century. This section explores how national innovation systems, science and technology policies, and governance mechanisms have influenced the continent's capacity to generate and sustain innovation.

3.1 National Innovation Systems (NIS) in Africa

The concept of a National Innovation System (NIS) provides a useful framework for understanding how countries generate, diffuse, and utilize knowledge for economic and social development. Originating from the works of scholars such as Christopher Freeman and Bengt-Åke Lundvall, the NIS framework emphasizes that innovation is not an isolated act of invention but a systemic process shaped by the interaction of firms, universities, research institutes, government agencies, and financial institutions (Freeman, 1987, 1988; Lundvall, 1992; Nelson, 1993). In essence, a country's innovative capacity depends not merely on individual talent or capital but on the institutional architecture and policy coherence that enable ideas to flow between knowledge producers and users.

In the African context, the NIS framework must be interpreted within the realities of historical dependency, institutional fragility, and developmental aspirations. Unlike in advanced economies, where innovation systems evolved organically through industrial revolutions, African systems are largely policy-driven constructs, often developed through international cooperation and donor influence. Many African nations have adopted STI (Science, Technology, and Innovation) policies inspired by OECD models, yet their translation into tangible outcomes has been uneven due to limited resources, coordination failures, and weak linkages between research and production sectors.

A distinguishing feature of African innovation systems is their hybrid character, a coexistence of formal and informal innovation networks. While national research institutions and ministries attempt to drive policy from the top, a substantial share of innovation takes place at the grassroots level: farmers developing adaptive techniques to respond to climate change, informal mechanics designing machinery from scrap materials, or local entrepreneurs creating digital applications for micro-trade and logistics. Thus, Africa's innovation systems often combine institutional experimentation and informal ingenuity, reflecting both structural constraints and creative adaptation.

Across the continent, one observes significant variations in the maturity and coherence of national innovation systems. Some countries, like South Africa and Kenya, have developed relatively structured frameworks linking academia, industry, and government, while others remain characterized by fragmented and underfunded institutions. The capacity to mobilize resources, integrate innovation into industrial policy, and align national priorities with global technological shifts varies widely. Understanding these variations provides critical insight into why certain African economies have been able to leverage technology for growth while others have lagged despite similar policy rhetoric.

3.2 Case Studies: South Africa, Kenya, Nigeria, and Egypt

3.2.1 South Africa's Institutional Depth and Policy Coherence

South Africa stands as the most advanced example of a functioning NIS in Africa. Building on a strong base of universities, public research councils, and industrial policy instruments, it has developed a relatively coherent innovation framework (National Planning Commission [NPC], 2012). Institutions such as the Department of Science and Innovation (DSI), the National Research Foundation (NRF), and the Technology Innovation Agency (TIA) play central roles in coordinating funding, research, and commercialization. The country's National System of Innovation White Paper outlines a clear vision linking science and technology to industrial competitiveness and social inclusion.

South Africa's innovation system is underpinned by strong university-industry linkages and sectoral specialization in areas like biotechnology, renewable energy, and mining technologies. The establishment of the Council for Scientific and Industrial Research (CSIR) and the Square Kilometre Array (SKA) project has positioned the country as a continental leader in research infrastructure. However, persistent inequality and skill mismatches continue to limit the diffusion of innovation beyond urban centers, creating an uneven geography of technological participation. Despite these challenges, South Africa's institutional capacity and policy coherence make it a benchmark for other African countries.

3.2.2 Kenya's Digital Innovation and Entrepreneurial Ecosystems

Kenya represents a contrasting model, one where innovation has flourished despite limited formal research infrastructure. Dubbed the "Silicon Savannah", Kenya's innovation success is rooted in the digital revolution, entrepreneurship, and adaptive policy support. The landmark mobile money platform M-Pesa transformed not only financial inclusion but also the global perception of African innovation. Building on this success, Kenya established the National Commission for Science, Technology, and Innovation (NACOSTI) and implemented the Vision 2030 Development Blueprint, integrating ICT and innovation into national planning (Government of Kenya, 2007).

Unlike South Africa's research-led model, Kenya's NIS thrives on bottom-up innovation. The proliferation of technology hubs such as iHub, Nailab, and Gearbox has created fertile ground for start-ups in fintech, agritech, and health tech. The government's openness to public-private partnerships and digital infrastructure investment (e.g., the Konza Technopolis project) reflects an ecosystem-oriented policy approach. Kenya's case illustrates how innovation systems in Africa can evolve around entrepreneurial networks rather than formal research hierarchies, aligning technology with local needs and inclusive growth.

3.2.3 Nigeria's Policy Ambition Amid Structural Constraints

Nigeria's innovation system reveals the tensions between policy ambition and implementation capacity. As Africa's largest economy, Nigeria possesses considerable human capital and a growing tech community, yet its innovation ecosystem remains constrained by institutional fragmentation, limited research funding, and weak industry-academia collaboration. The National Policy on Science, Technology, and Innovation (2022–2030) outlines an ambitious agenda for technological self-reliance and industrial diversification, but coordination among federal and state institutions remains inconsistent (Government of Nigeria, 2022).

Nevertheless, Nigeria has witnessed remarkable growth in private-sector-driven innovation, particularly in fintech and the creative industries. Start-ups such as Flutterwave, Paystack, and Andela have attracted global attention and investment, positioning Nigeria as a continental leader in digital entrepreneurship. The government's National Information Technology Development Agency (NITDA) has also introduced supportive frameworks for ICT adoption and startup incubation. Yet, infrastructural bottlenecks, policy inconsistency, and limited intellectual property protection continue to hinder the evolution of a fully integrated NIS. Nigeria's experience underscores the need to strengthen institutional coordination and incentivize research-industry collaboration as key drivers of innovation-led growth.

3.2.4 Egypt's Science-Led Transformation and Regional Leadership

Egypt presents a hybrid model combining state-led innovation planning with growing private sector participation. With a long tradition of higher education and scientific research, Egypt has leveraged its institutional heritage to build one of Africa's most structured innovation systems. The Ministry of Higher Education and Scientific Research (MHESR), through agencies such as the Academy of Scientific Research and Technology (ASRT), coordinates national R&D efforts and fosters collaboration among universities, research centers, and industries.

Egypt's Vision 2030 prioritizes knowledge economy transformation, with significant investment in ICT, renewable energy, and biotechnology. The establishment of Smart Villages, technology parks, and incubators has catalyzed innovation, especially among youth-led enterprises. Egypt's participation in regional initiatives like STISA-2024 and cooperation with the African Union Commission further strengthen its position as a continental innovation leader. The challenge, however, lies in translating scientific output into marketable technologies and ensuring equitable participation across social groups.

The four case studies reveal that while African countries share common challenges—such as limited R&D investment, weak innovation linkages, and regulatory hurdles—they also exhibit distinct institutional logics. South Africa demonstrates the power of coordinated institutions; Kenya showcases entrepreneurial dynamism; Nigeria reflects the promise and pitfalls of scale without strong governance; and Egypt highlights the value of scientific tradition in building innovation capacity. Together, they affirm that Africa's National Innovation Systems are evolving mosaics, each shaped by history, policy choices, and socioeconomic realities. The lesson is clear: there is no one-size-fits-all model for innovation in Africa, but success depends on building context-sensitive systems that bridge the divide between knowledge creation and productive application.

3.3 Science, Technology, and Innovation (STI) Policies

The evolution of Science, Technology, and Innovation (STI) policies in Africa reflects the continent's broader political and economic transitions from state-centered planning after independence to market-oriented reforms and, more recently, hybrid models emphasizing partnerships, entrepreneurship, and knowledge economies. In the immediate post-independence period (1960s–1970s), most African governments viewed technology as a lever for rapid industrialization and modernization. Influenced by developmental state ideologies, they adopted state-led strategies characterized by heavy public investment in industrial projects, research institutes, and parastatal enterprises. The state was envisioned as the central actor responsible for technological acquisition, diffusion, and coordination.

However, these early efforts were largely import-substitution industrialization (ISI) models that prioritized physical infrastructure and imported machinery over indigenous technological capacity. Most countries lacked the scientific infrastructure, human capital, and institutional coherence required to sustain innovation locally. Consequently, many state-led projects became economically unsustainable, particularly after the global oil crises and debt burdens of the 1970s and 1980s. The subsequent Structural Adjustment Programs (SAPs), imposed by the International Monetary Fund (IMF) and the World Bank in the 1980s, ushered in a wave of market liberalization and privatization, which fundamentally altered Africa's innovation policy landscape.

During the 1990s, the policy focus shifted from direct state intervention to creating enabling environments for private investment, foreign technology inflows, and trade liberalization. Governments were encouraged to minimize their role in production and instead focus on regulation, infrastructure, and education. This period saw a decline in public research funding but an increase in partnerships with foreign donors and multilateral organizations. STI was often subordinated to macroeconomic stabilization rather than treated as a central pillar of development policy.

The early 2000s marked a renewed recognition of STI as a driver of sustainable development. The emergence of the knowledge economy paradigm, trade liberalization through international protocols and bilateral agreements, and the influence of global development frameworks such as the Millennium Development Goals (MDGs) and later the Sustainable Development Goals (SDGs) reinvigorated the policy discourse around science and technology. African governments began to adopt comprehensive STI policies, integrating innovation into national development plans. These policies emphasized entrepreneurship, digital transformation, and capacity building rather than merely technology transfer.

Today, many African nations operate within a mixed or hybrid policy model balancing market-driven innovation ecosystems with state support mechanisms for research and strategic sectors. Governments play a facilitating role through innovation funds, tax incentives, and digital infrastructure, while the private sector and civil society drive technological application and diffusion. For instance, Kenya's Vision 2030, South Africa's National Development Plan, and Nigeria's STI Policy (2022–2030) each reflect this balanced approach, combining top-down planning with bottom-up innovation ecosystems.

3.4 Governance, Regulation, and Intellectual Property Regimes

The governance of innovation in Africa is influenced by law, policy, and institutional coordination. Effective innovation governance requires legal frameworks that not only protect intellectual property but also facilitate technology transfer, entrepreneurship, and fair competition. However, Africa's experience reveals a system marked by uneven legal development, fragmented implementation, and competing policy objectives. While most African countries have enacted Science, Technology, and Innovation (STI) or intellectual property (IP) laws, their operationalization often lags due to limited institutional capacity, weak enforcement mechanisms, and low public awareness of intellectual property rights.

The legal foundations for innovation governance in Africa draw from both national legislation and international conventions. Most African states are signatories to global instruments such as the World Intellectual Property Organization (WIPO) Convention, the Paris Convention for the Protection of Industrial Property, and the Berne Convention for the Protection of Literary and Artistic Works. These commitments provide the structural basis for regulating patents, trademarks, copyrights, and industrial designs. Furthermore, Africa hosts regional IP organizations, most notably, the African Regional Intellectual Property Organization (ARIPO) and the Organisation Africaine de la Propriété Intellectuelle (OAPI), which harmonize IP laws among member countries, reduce administrative burdens, and promote cross-border enforcement.

In recent years, the African Continental Free Trade Area (AfCFTA) has introduced new momentum for legal and regulatory convergence (African Continental Free Trade Area Secretariat, 2023). Its Protocol on Intellectual Property Rights, once fully operational, aims to streamline IP protection across borders, stimulate investment, and enhance the transfer of technology among member states. Complementing this, the African Union Commission (AUC) and United Nations Economic Commission for Africa (UNECA) have advanced policy dialogues on creating an African Common Intellectual Property Framework, emphasizing inclusive innovation and equitable access to knowledge (African Union Commission [AUC], 2015; UNECA, 2021).

However, translating these frameworks into practical mechanisms for innovation and technology transfer remains challenging. Many African firms and research institutions face significant barriers to accessing global technologies due to restrictive licensing terms, high costs, and weak domestic absorptive capacity. Technology transfer agreements are often embedded within foreign direct investment (FDI) contracts or development aid packages, where asymmetrical power dynamics limit local bargaining positions. As a result, technology inflows may reinforce dependency rather than build endogenous innovation capacity.

A related governance issue is the fragmentation of institutional mandates. Ministries of science and technology, education, trade, and justice often operate in silos, leading to policy incoherence. For example, while one agency promotes open innovation and collaboration, another may impose restrictive IP enforcement measures that discourage sharing. Moreover, many African countries lack specialized IP courts or trained judges capable of interpreting complex patent disputes, which undermines investor confidence and slows innovation diffusion. Strengthening governance, therefore, requires not only legislative reforms but also institutional alignment, judicial capacity building, and greater participation of private and informal sector innovators in policy design.

4 Business Innovation and the Digital Economy

Africa's economic transformation is increasingly being powered by digital technologies and entrepreneurial ingenuity. Across the continent, a new generation of innovators is harnessing mobile connectivity, data analytics, and emerging technologies to create solutions for finance, agriculture, health, and education. The digital economy has become more than a sector; it is the foundation of a new business culture defined by creativity, agility, and inclusion. This section explores how African entrepreneurs, start-ups, and digital ecosystems are driving growth, attracting investment, and positioning the continent as a vibrant player in the global innovation landscape.

4.1 Role of Venture Capital, Incubators, and Accelerators

The rapid growth of Africa's start-up ecosystem has been accompanied by a parallel expansion in venture capital (VC) activity and support infrastructure. Historically, African entrepreneurs faced enormous barriers to finance due to underdeveloped capital markets, risk-averse banking systems, and limited angel investment. Over the past decade, however, the narrative has shifted dramatically. Venture funding in Africa has grown from less than $500 million in 2015 to over $6 billion by 2024, concentrated primarily in fintech, e-commerce, logistics, and climate technology.

The venture capital ecosystem in Africa is increasingly global, with active participation from investors in North America, Europe, and Asia. Funds such as Tekedia Capital, Partech Africa, TLcom Capital, Helios Investment Partners, Future Africa, and Chandaria Capital have become central players in financing early-stage ventures. International technology giants like Google, Meta, Microsoft, and Amazon Web Services have also expanded their accelerator programmes across the continent, providing mentorship, technical support, and networking opportunities. Local investors and diaspora funds have joined the ecosystem as well, recognizing Africa's potential as the next frontier of digital innovation.

In parallel, incubators and accelerators have proliferated as critical components of the entrepreneurial ecosystem. These institutions provide training, seed funding, co-working spaces, and mentorship that help start-ups refine their business models and attract investment. Nairobi's iHub, Cape Town's Workshop17, and Lagos's CcHub (Co-Creation Hub) are among the most influential, having supported hundreds of start-ups that have grown into regional or global players. The Tony Elumelu Foundation Entrepreneurship Programme has further democratized access to entrepreneurship support, reaching over 15,000 African start-ups across all 54 countries.

Governments and development agencies have also played instrumental roles in expanding innovation support networks. The World Bank's Digital Economy for Africa (DE4A) initiative, the African Development Bank's Boost Africa Program, and the UNDP's Timbuktoo Project have injected technical and financial support into innovation ecosystems, promoting sustainable entrepreneurship across the continent. Moreover, several African countries now operate sovereign innovation funds such as Rwanda's Innovation Fund and South Africa's Small Enterprise Finance Agency (SEFA) to catalyze local venture capital markets and bridge financing gaps for early-stage firms.

Despite these encouraging trends, challenges persist. Access to funding remains heavily concentrated geographically: over 70% of venture capital still flows to Nigeria, Kenya, South Africa, and Egypt, leaving other countries underfunded. Furthermore, many investors remain hesitant about long-term returns due to currency volatility, regulatory uncertainty, and weak exit markets. These constraints underscore the need for policy reforms that de-risk innovation investment, enhance credit infrastructure, and support regional capital mobility under frameworks such as the AfCFTA.

Importantly, the expansion of venture capital and accelerator ecosystems has begun to reshape entrepreneurial culture and aspiration across Africa. Young people increasingly view entrepreneurship not as a fallback option but as a legitimate career path and driver of national transformation. This cultural shift, combined with the democratization of digital tools, has produced a new generation of innovators who are confident, globally connected, and locally grounded. Their work embodies the continent's capacity for inclusive, frugal, and context-driven innovation, which seeks to solve societal challenges rather than merely replicate foreign business models.

4.2 Industry 4.0 and Emerging Technologies

The advent of Industry 4.0, characterized by the fusion of digital, biological, and physical technologies, has opened new frontiers for Africa's development. Although often viewed as a latecomer to earlier industrial revolutions, Africa is increasingly positioning itself to leapfrog traditional pathways by strategically adopting emerging technologies. AI, blockchain, fintech, agritech, and renewable energy are at the forefront of this transformation, reshaping how African societies produce, trade, govern, and live.

Artificial Intelligence (AI) is gradually moving from a niche academic field to a practical tool for solving real-world African problems. Across the continent, AI is being applied in healthcare for diagnostics and disease prediction, in agriculture for crop monitoring and yield optimization, in textile manufacturing for warehousing and for carbon neutrality and in financial services for credit scoring and fraud detection. Initiatives such as Data Science Nigeria, Deep Learning Indaba, and AI4D Africa (Artificial Intelligence for Development in Africa) are training a new generation of data scientists and promoting ethical AI adoption tailored to African contexts. In Rwanda and Tunisia, AI is being integrated into e-governance and public service delivery, while South Africa is exploring AI applications in mining safety, logistics, and urban planning. Mauritius and Seychelles, both African Small Island Developing States (SIDS), are also advancing in the AI race. Mauritius leverages AI in its textile industry to optimize energy use and minimize environmental impact, while Seychelles is implementing an AI-powered customs system at Port Victoria. Despite data limitations and infrastructure challenges, Africa's AI movement signals a growing recognition that intelligent automation and data analytics are not luxuries but essential tools for inclusive, evidence-based development.

Blockchain technology offers another transformative potential. Originally associated with cryptocurrencies, blockchain has matured into a broader infrastructure for trust, transparency, and decentralization. In Africa, it is being used to authenticate land titles (as seen in Ghana's Bitland project), track agricultural supply chains, improve public procurement transparency, and facilitate cross-border payments. Governments and start-ups alike are experimenting with blockchain-based identity systems to enhance governance integrity and reduce fraud. For instance, Kenya's Ministry of ICT has explored blockchain for managing public records, while Nigeria is piloting its eNaira, a central bank digital currency designed to promote financial inclusion. The continent's reliance on remittances and informal trade makes blockchain particularly relevant as a tool to reduce transaction costs and empower individuals excluded from formal financial systems. Yet, for blockchain to reach its full potential, regulatory clarity, energy-efficient infrastructure, and public trust must deepen.

The fintech revolution is perhaps Africa's most visible expression of technological disruption. With over 500 million unbanked adults and limited access to traditional banking infrastructure, fintech has emerged as a democratizing force in finance. Companies such as M-Pesa (Kenya), Flutterwave (Nigeria), Chipper Cash (Uganda), and Jumo (South Africa) have redefined financial inclusion through mobile payments, digital lending, and cross-border transfers. The expansion of fintech is not only improving financial access but also driving entrepreneurship, enabling micro-insurance, and expanding access to credit for small businesses. Regulators, initially cautious, are now catching up by introducing regulatory sandboxes, controlled environments that allow innovation without undermining financial stability. As fintech becomes intertwined with AI and blockchain, Africa's financial systems are evolving into digital ecosystems, creating entirely new value chains and opportunities for regional integration under the AfCFTA.

Renewable energy technologies are also reshaping Africa's development landscape by addressing one of its most persistent bottlenecks, i.e., energy access. With over 600 million Africans still lacking electricity, the traditional grid model is giving way to decentralized renewable systems powered by solar, wind, and biomass. Start-ups like M-KOPA (Kenya), Bboxx (Rwanda), and Lumos (Nigeria) are pioneering pay-as-you-go solar models that bring clean energy to rural households through mobile payments (International Energy Agency [IEA], 2022). Governments, too, are increasingly integrating renewables into national energy strategies, as seen in Morocco's Noor Solar Complex, one of the world's largest concentrated solar plants, and South Africa's Renewable Energy Independent Power Producer Procurement Programme (REIPPPP). The renewable energy sector exemplifies how technological innovation can simultaneously tackle environmental sustainability, job creation, and social equity pillars of Africa's green transformation agenda.

In parallel, agritech innovations are redefining Africa's agricultural landscape, which remains the primary livelihood for nearly 60% of the population. Digital platforms such as Twiga Foods (Kenya), Farmcrowdy (Nigeria), and Hello Tractor are connecting smallholder farmers to markets, credit, and mechanization services. Drones and satellite imaging are being used to monitor soil conditions, predict rainfall patterns, and combat pests. Start-ups are leveraging data analytics to optimize crop yields and reduce post-harvest losses, while governments are increasingly adopting e-extension systems to support farmers in remote regions. The combination of IoT sensors, mobile platforms, and AI-driven analytics is creating a more data-informed agricultural sector, boosting productivity while enhancing resilience to climate change.

Together, these technologies are not operating in isolation but forming an interconnected ecosystem that underpins Africa's digital economy. AI supports fintech credit models, blockchain strengthens supply chain transparency, and renewable energy powers the data centers and rural connectivity essential for digital inclusion. What differentiates Africa's adoption of Industry 4.0 from that of other regions is its problem-solving orientation, where technologies are not just adopted for prestige but for addressing urgent societal challenges such as poverty, exclusion, and environmental degradation.

4.3 Private Sector, Diaspora, and Global Partnerships

The private sector has become the cornerstone of Africa's technological and innovation renaissance. While government policies and donor programs have laid the groundwork for science, technology, and innovation development, it is private enterprise supported by transnational networks and capital flows that has fuelled the rapid expansion of Africa's digital and entrepreneurial ecosystems. The continent's innovation story is no longer solely about aid or public investment; it is increasingly a narrative of venture-backed start-ups, multinational collaborations, and diaspora-led enterprises that link African ingenuity to global markets.

Foreign Direct Investment (FDI) has played a central role in catalyzing technological growth. Over the past decade, global technology companies have deepened their footprint in Africa, seeking both emerging markets and innovation opportunities. Firms such as Google, Microsoft, Meta (formerly Facebook), Amazon Web Services, Huawei, and Orange have invested in digital infrastructure, cloud computing, and training programs across the continent. These investments are not purely philanthropic; they represent strategic engagement with Africa's fast-growing digital consumer base, expected to exceed one billion users by 2030. The establishment of data centers by Google in South Africa, Microsoft in Kenya, and Huawei's investment in 5G networks exemplify how Africa's digital economy is now seen as a frontier of global innovation rather than a periphery of consumption.

FDI has also become increasingly diversified and collaborative. Beyond traditional Western investors, Asian and Middle Eastern economies, notably China, India, Singapore, and the United Arab Emirates, have emerged as pivotal players. China's Digital Silk Road initiative, a component of the Belt and Road strategy, has financed fiber-optic networks, smart city projects, and e-commerce logistics hubs in countries like Ethiopia, Kenya, and Nigeria. Indian firms such as Bharti Airtel and Tech Mahindra have invested in ICT infrastructure and digital financial services, while Gulf-based investors have financed renewable energy and agritech ventures. These investments bring not only capital but also technical expertise and global market access, enabling African start-ups to scale regionally and internationally.

The role of diaspora networks has been particularly transformative. Africa's global diaspora, estimated at over 150 million people, acts as a bridge connecting the continent to advanced innovation ecosystems in North America, Europe, and Asia. Diaspora entrepreneurs and professionals contribute to Africa's innovation ecosystem in three key ways: capital infusion, knowledge transfer, and network building. Through remittances, angel investment, and venture funds such as Future Africa, Afropreneur Fund, and Golden Palm Investments, diaspora actors finance early-stage start-ups that often struggle to secure domestic funding. Equally important, they bring managerial expertise, global exposure, and mentorship that help African founders navigate the complexities of scaling in competitive markets.

The success of companies like Andela, Flutterwave, and Kobo360 illustrates this symbiosis. Each was co-founded or supported by diaspora professionals who combined international experience with local insight. Moreover, diaspora-led innovation initiatives such as The African Diaspora Network (ADN) in Silicon Valley and Homestrings in the UK create platforms for structured investment in African innovation. The African Union's Diaspora Division and AU–CIDO (Citizens and Diaspora Organizations Directorate) are also institutionalizing diaspora engagement as part of the continental development agenda under Agenda 2063, framing the diaspora as Africa's "sixth region."

Yet, while transnational networks have deepened Africa's integration into the global innovation economy, they also expose the continent to new dependencies and vulnerabilities. The dominance of foreign capital in the start-up ecosystem raises concerns about profit repatriation, data sovereignty, and strategic autonomy. For sustainable innovation growth, African countries must not only attract FDI but also negotiate value retention mechanisms to ensure that technology transfer, local ownership, and skill development accompany foreign partnerships. In this respect, balancing openness with strategic regulation will be crucial for ensuring that Africa's innovation ecosystems serve continental priorities rather than external interests.

While transnational capital is reshaping Africa's innovation landscape, the public–private partnership (PPP) model remains the linchpin for scaling technological solutions to national and regional levels. PPPs bridge the resource and capacity gap between government institutions and private investors, allowing each to play complementary roles. Governments provide policy frameworks, infrastructure, and regulatory stability; private actors bring agility, innovation, and financial capital.

The success of PPPs across Africa demonstrates how coordinated action can transform technology from an elite privilege into a public good.

In infrastructure and connectivity, PPPs have been instrumental. The construction of the East African Submarine Cable System (EASSy) and the West African Cable System (WACS), jointly financed by governments, telecom operators, and multilateral agencies, dramatically reduced internet costs and improved broadband penetration. In Kenya, collaboration between the government and private telcos led to the National Optical Fibre Backbone Infrastructure (NOFBI), connecting previously underserved regions. Similarly, South Africa's Renewable Energy Independent Power Producer Procurement Programme (REIPPPP) showcases how government procurement combined with private investment can rapidly expand clean energy capacity while creating jobs and local supply chains.

In the realm of digital entrepreneurship, PPPs are supporting innovation ecosystems that combine state facilitation with private execution. Rwanda's Kigali Innovation City and Egypt's Smart Village Cairo represent successful models where government planning, corporate investment, and academic research intersect to drive innovation. Nigeria's National Digital Economy Policy (2020–2030) also hinges on PPPs to promote broadband infrastructure, digital skills, and e-government services. These initiatives highlight a growing recognition that governments alone cannot drive technological transformation; rather, sustainable progress requires a coalition of actors aligned by shared incentives and accountability.

Beyond domestic partnerships, Africa is increasingly engaging in South–South collaboration, cooperation among developing nations to share technology, knowledge, and best practices. This form of collaboration provides a viable alternative to the traditional North–South aid paradigm, emphasizing mutual learning and co-development. China, India, Brazil, and Turkey have been at the forefront of this approach, supporting African innovation through technology parks, training programs, and joint ventures. For example, India's Pan-African e-Network Project has connected African universities and hospitals to Indian institutions via tele-education and telemedicine platforms. Similarly, Brazil has supported agritech research through partnerships with the Embrapa (Brazilian Agricultural Research Corporation) and African agricultural research centers.

Inter-African collaboration under the African Continental Free Trade Area (AfCFTA) and Smart Africa Alliance further strengthens this trend. By harmonizing digital regulations, promoting data interoperability, and facilitating innovation clusters across borders, these frameworks encourage African countries to build collective digital sovereignty. The African Development Bank (AfDB) and the Economic Commission for Africa (ECA) are actively funding cross-border innovation corridors that integrate SMEs, universities, and start-ups from multiple countries.

However, sustaining PPPs and South–South partnerships requires attention to governance and equity. Many partnerships face obstacles such as bureaucratic delays, unclear legal frameworks, and limited project monitoring. Moreover, smaller domestic firms often struggle to compete with multinational corporations in PPP arrangements. To ensure inclusivity, African governments must create transparent tendering processes, prioritize local content requirements, and embed technology transfer clauses in all major contracts.

5 Socioeconomic and Developmental Implications

The technological transformation unfolding across Africa is not only reshaping industries and governance but also redefining the continent's socioeconomic fabric. As digital technologies penetrate every sector from finance and agriculture to education and healthcare, they are altering how Africans work, learn, and interact. Yet, these changes are complex and multidimensional; they create new opportunities for inclusion and empowerment while simultaneously exposing deep structural inequalities. Understanding the socioeconomic and developmental implications of this transformation requires examining how technology influences productivity, employment, inclusion, and inequality within Africa's broader developmental context.

For decades, Africa's economic growth has been driven largely by commodity exports and low-value services. Technological innovation offers a pathway to structural transformation, enabling economies to move from resource dependence to knowledge- and service-based production. Emerging technologies, especially digital platforms, automation, and data analytics, are enhancing efficiency in agriculture, manufacturing, logistics, and retail.

In agriculture, digital platforms such as Twiga Foods, Farmcrowdy, and AgriPredict are improving supply chain coordination, reducing waste, and increasing farmer incomes. In manufacturing, additive manufacturing (3D printing) and Internet of Things (IoT) technologies are beginning to transform small-scale production systems, enabling local firms to compete globally without large-scale industrial infrastructure. Similarly, digital logistics firms like Kobo360 and Lori Systems are reducing inefficiencies in transport networks, which historically increased the cost of intra-African trade.

At a macro level, these technological improvements are raising productivity by reducing transaction costs and unlocking new sources of comparative advantage (UNECA, 2021). For instance, Rwanda's use of drones for medical deliveries has not only improved health outcomes but also lowered distribution costs by more than 50% in rural areas. The integration of data analytics into public administration, e-government systems, digital registries, and smart tax platforms is enhancing fiscal transparency and resource mobilization. These innovations, when scaled, could accelerate Africa's industrialization by aligning technological progress with inclusive economic policy.

5.1 Employment, Skills, and the Future of Work

The rise of digital technologies presents a paradox for Africa's labor markets. On the one hand, automation and artificial intelligence threaten traditional jobs, particularly in low-skill, routine-based sectors. On the other hand, the digital economy is creating entirely new categories of employment, from app-based logistics and digital marketing to coding, design, and remote freelancing. The World Bank estimates that Africa's digital economy could add $180 billion to GDP by 2025 and $715 billion to GDP by 2050, generating millions of jobs across the value chain.

However, these opportunities depend heavily on human capital development. Africa's educational systems must evolve from rote learning to problem-solving, creativity, and digital literacy. Countries such as Kenya, Mauritius, and Rwanda have begun integrating coding and computational thinking into school curricula, while universities are expanding programs in data science, artificial intelligence, and digital business. Non-traditional learning platforms such as Andela Learning Community, ALX Africa, and Moringa School are also bridging the digital skills gap through intensive, industry-aligned training.

Informal employment, which accounts for over 80% of Africa's workforce, is undergoing digital transformation as well. Mobile platforms like Jumia, Glovo, and Bolt have expanded gig-based work, providing flexible income opportunities while raising concerns about precarity and labor protection. The challenge for policymakers is to ensure that digital work remains dignified, fair, and inclusive by creating new frameworks for social protection, digital taxation, and collective bargaining suited to the realities of platform economies.

5.2 Digital Inclusion, Inequality, and the Urban–Rural Divide

While technology has the potential to democratize opportunity, it can also exacerbate inequality if access remains uneven. A persistent digital divide, along socioeconomic, geographic, and gender lines, continues to shape who benefits from technological progress in Africa. Urban centers such as Nairobi, Lagos, and Johannesburg enjoy high-speed internet, innovation hubs, and venture capital access, while many rural areas still struggle with unreliable electricity and limited connectivity.

Gender disparities in digital access are particularly concerning. According to the GSMA Mobile Gender Gap Report (2024), African women are 25% less likely than men to use mobile internet, largely due to affordability, literacy, and social norms. This exclusion restricts women's participation in digital entrepreneurship, online learning, and financial inclusion. Addressing these disparities requires targeted interventions such as subsidized data plans, digital literacy programs for women and youth, and the inclusion of gender equality objectives in national ICT and innovation policies.

Equally critical is infrastructure equity. The expansion of fiber-optic networks, rural broadband initiatives, and satellite-based internet, such as Starlink and Avanti Communications, is beginning to bridge connectivity gaps. However, equitable development demands more than physical access; it requires affordability, relevant digital content, and institutional support for marginalized communities. Without these, the digital revolution could entrench new forms of exclusion mirroring old socioeconomic hierarchies.

5.3 Social Mobility, Entrepreneurship, and Empowerment

Technology has emerged as a powerful engine of social mobility and empowerment across Africa. The democratization of digital tools such as smartphones, e-commerce platforms, and mobile payments has enabled individuals and micro-enterprises to access markets, finance, and knowledge previously out of reach. For instance, artisans in Ghana, fashion entrepreneurs in Nigeria, and small-scale traders in Kenya now reach global customers through platforms like Etsy, Jumia, and Shopify. Fintech services have given informal businesses credit histories and digital transaction records, enhancing their legitimacy and growth potential.

At the community level, technology is amplifying voices and strengthening civic engagement. Social media and digital platforms have enabled citizens to organize movements, monitor governance, and demand accountability. The growth of civic tech such as BudgIT in Nigeria and Code for Africa illustrates how data transparency tools can foster participatory governance. Moreover, in post-conflict and fragile states, mobile-based communication platforms have improved access to humanitarian aid and reconnected displaced populations.

Among youth, technology has become synonymous with agency and aspiration. The rise of innovation hubs, hackathons, and entrepreneurship competitions like Innovate@Wigwe, HiiL Justice Accelerator, and Orange Fab Africa has created spaces for creative problem-solving and collaboration. These platforms are redefining African youth identities from job seekers to job creators, linking innovation to dignity and self-determination.

5.4 Balancing Growth with Equity and Sustainability

While technology-driven growth is reshaping Africa's economic landscape, it also raises concerns about sustainability and equity. The extraction of minerals for digital devices, electronic waste management, and the carbon footprint of data centers pose environmental challenges that demand policy attention. Likewise, the dominance of a few digital giants, both foreign and domestic, creates risks of data monopolies and market concentration, which could stifle competition and innovation.

Policymakers must therefore pursue a "just digital transition", ensuring that the gains from technological progress are widely distributed and environmentally sustainable. This includes regulating platform economies to protect workers, encouraging open innovation to prevent knowledge monopolies, and integrating digitalization into national sustainability agendas. Frameworks such as the African Green Digital Compact (AGDC), under the African Union's Agenda 2063, provide pathways for aligning digital transformation with climate resilience and social inclusion.

6 Challenges and Structural Constraints

Despite the promising momentum of Africa's digital transformation, the path to a fully integrated and self-sustaining innovation ecosystem remains fraught with structural barriers. These constraints, ranging from infrastructural deficits and financial gaps to weak policy coherence, limited absorptive capacity, and dependence on foreign technology, undermine the continent's ability to translate innovation potential into broad-based development. Understanding these challenges is essential for designing strategic interventions that ensure Africa's technological revolution is inclusive, autonomous, and sustainable.

6.1 Infrastructure Deficits, Financing Gaps, and Brain Drain

Africa's most enduring constraint on technological advancement is its infrastructure deficit. Reliable power, broadband connectivity, transport networks, and data infrastructure form the backbone of any modern innovation system, yet in many African countries these remain severely underdeveloped. Approximately 600 million Africans still lack access to electricity, and unreliable power supply continues to cripple businesses and research institutions (International Energy Agency, 2022). The absence of stable energy sources hinders the operation of laboratories, data centers, and digital start-ups that rely on uninterrupted connectivity and computing capacity.

Broadband infrastructure has expanded rapidly in recent years through submarine cables, fiber-optic investments, and satellite initiatives. However, coverage and affordability remain uneven. Urban centers enjoy high-speed connectivity, while vast rural regions remain digitally isolated. The cost of internet access, often more than 5% of average monthly income, remains prohibitive for many citizens. This infrastructure imbalance not only deepens social inequality but also limits the scalability of innovation across the continent.

Closely tied to infrastructure challenges are financing gaps. Despite rising venture capital inflows, the majority of African start-ups and research institutions struggle to access affordable capital. Financial systems remain underdeveloped, risk-averse, and skewed toward short-term lending. Traditional banks are hesitant to fund innovation due to its intangible nature and uncertain returns. Public funding for research and development is chronically low, less than 1% of GDP in most countries, undermining the sustainability of science and technology institutions.

Furthermore, the African financial ecosystem lacks a strong secondary market for innovation financing. Venture capital remains concentrated in a handful of countries like South Africa, Kenya, Nigeria, and Egypt, while local angel investor networks are nascent elsewhere. Development finance institutions and innovation funds exist but often lack autonomy and efficiency. The result is a financing paradox where there is abundant entrepreneurial talent but insufficient long-term investment to transform ideas into scalable enterprises.

Compounding these economic and infrastructural constraints is the persistent brain drain of skilled professionals. Africa heavily subsidizes higher education, yet many of its scientists, engineers, and innovators emigrate in search of better opportunities abroad. According to the African Development Bank (2023), nearly 30% of African-born professionals live and work outside the continent. This loss deprives Africa of technical expertise, mentorship, and institutional continuity. While digital connectivity now allows for "virtual return", where diaspora experts contribute remotely, reversing the brain drain requires structural reforms that make research careers viable at home, namely, competitive remuneration, functional laboratories, merit-based funding, and political stability.

6.2 Weak Policy Coherence and Limited Absorptive Capacity

Another critical barrier lies in the fragmented and inconsistent policy environment governing science, technology, and innovation. While most African countries have national STI strategies, these are often disjointed, poorly funded, or disconnected from broader economic planning. Ministries of education, trade, ICT, and industry frequently operate in isolation, pursuing parallel agendas without coordination. The absence of clear monitoring and evaluation frameworks leads to policy fatigue; numerous documents with limited implementation.

Moreover, the short political cycles and frequent ministerial reshuffles that characterize many African governments disrupt continuity in innovation policy. Long-term research programs are abandoned with changes in administration, and institutional reforms stall due to shifting political priorities. The lack of data-driven policy design further compounds the problem: few countries maintain robust innovation statistics or performance indicators, making it difficult to assess outcomes or learn from past interventions.

Equally pressing is the issue of limited absorptive capacity, the ability of domestic institutions and firms to internalize, adapt, and utilize new technologies. Many African businesses remain small, informal, and technologically underequipped. They often lack the managerial skills, R&D capacity, and networks required to translate innovation into productivity gains. This limitation extends to universities and research institutes, where bureaucratic inertia, underfunding, and weak industry linkages hinder technology transfer. The result is a "knowledge disconnect," where innovation generated in academic or policy circles fails to reach the marketplace.

To address this, Africa must build institutional learning ecosystems that connect research, enterprise, and government in a dynamic feedback loop. Countries like Rwanda and Mauritius have begun experimenting with performance-based funding, innovation councils, and knowledge transfer offices that reward measurable outcomes rather than bureaucratic compliance. However, widespread replication of such models requires political will, capacity building, and sustained investment in institutional reform.

6.3 Dependence on Foreign Technology and the Risk of Digital Colonialism

While globalization has opened Africa to new technologies and markets, it has also reinforced patterns of technological dependency that mirror older forms of economic colonialism. The majority of Africa's digital infrastructure, software, and data systems are owned, designed, or controlled by foreign corporations. From cloud storage to social media platforms, African users generate massive volumes of data that are stored offshore and monetized elsewhere, often with limited regulatory oversight. This dynamic has given rise to the term "digital colonialism", a twenty-first-century phenomenon where data and digital resources become the new commodities of extraction.

The dominance of global technology firms Google, Meta, Amazon, Huawei, and others has created both opportunities and vulnerabilities. On one hand, these firms bring investment, innovation, and global connectivity; on the other, they entrench asymmetric power relations where African states have limited control over data governance, taxation, or content moderation. Algorithms designed in Silicon Valley or Shenzhen increasingly shape African economies, politics, and culture, raising profound ethical and sovereignty concerns.

Africa's dependence also extends to hardware and critical infrastructure. Nearly all of the continent's servers, semiconductors, and telecom equipment are imported. Local manufacturing capacity in these areas remains negligible, leaving African economies exposed to global supply chain disruptions and foreign policy shifts. The absence of domestic technological sovereignty constrains Africa's ability to innovate independently or protect its digital infrastructure from external manipulation.

To mitigate this risk, several African institutions and regional bodies are advancing the agenda for digital sovereignty and data governance. The African Union's Data Policy Framework (2022) seeks to harmonize national regulations, promote local data storage, and ensure ethical AI practices. Similarly, the Smart Africa Alliance is working toward regional data centers and digital ID interoperability, while countries like Nigeria, Kenya, and South Africa are enacting local data protection laws inspired by the EU's GDPR. These efforts are crucial steps toward reclaiming control over Africa's digital future, yet they remain fragmented and under-resourced.

The danger of digital colonialism is not merely economic; it is existential. If left unchecked, Africa risks becoming a perpetual technology consumer, dependent on external actors for its digital infrastructure, algorithms, and even knowledge frameworks. This would entrench a form of dependency that undermines both innovation sovereignty and democratic accountability. The challenge, therefore, is to cultivate homegrown technological ecosystems that integrate global partnerships without forfeiting agency or autonomy.

7 Pathways for the Future

Africa's technological journey stands at a decisive juncture, rich with promise but burdened by structural fragilities. The future depends on whether the continent can consolidate recent gains and transform isolated innovation successes into sustainable, systemic progress. Building an inclusive, knowledge-based economy requires deliberate action across education, research, governance, and market integration. The following pathways offer strategic directions for achieving that vision.

7.1 Strengthening Education–Innovation Linkages and Research Capacity

A robust innovation ecosystem begins with education that inspires inquiry and application rather than rote memorization. For decades, Africa's education systems have excelled at producing graduates but struggled to nurture problem-solvers and inventors. Bridging this gap demands deep curricular reform from primary to tertiary levels to promote creativity, digital literacy, and interdisciplinary thinking.

At the tertiary level, universities must evolve from teaching institutions into research and innovation hubs that collaborate actively with industry and government. Establishing technology transfer offices, university–industry liaison units, and innovation clusters within campuses can accelerate the commercialization of research outputs. Countries like South Africa and Egypt already host strong university–industry linkages, while Kenya and Rwanda are experimenting with innovation-driven higher education models. Expanding these efforts continent-wide would ensure that academic research translates into industrial solutions and entrepreneurial ventures.

Research funding mechanisms must also be reimagined. Governments should institutionalize national research endowments and innovation challenge funds that provide stable, merit-based financing for scientists and technologists. Continental coordination through bodies like the African Academy of Sciences (AAS) and its likes from other disciplines can facilitate cross-border research collaborations and reduce duplication of effort. Furthermore, partnerships with the private sector should include co-funded R&D programs that align academic research with real-world market needs.

Finally, investment in STEM education, vocational training, and digital skills should be prioritized. While elite universities play a key role, technical and vocational institutions must be empowered to produce technicians, data analysts, and coders who bridge theory and practice. Without this foundation, Africa risks having pockets of excellence in research but a weak human capital base for technological diffusion.

7.2 *Building Regional Value Chains and Cross-Border Technology Markets*

No single African nation can independently sustain a globally competitive innovation economy. Fragmented markets, small domestic demand, and limited industrial diversification make regional integration indispensable. The African Continental Free Trade Area (AfCFTA) provides the framework for transforming Africa's 55 economies into a single digital and innovation market, a platform for scale, investment, and knowledge mobility (African Continental Free Trade Area Secretariat, 2023).

The next phase of Africa's transformation must therefore focus on building regional value chains in technology-intensive sectors such as renewable energy, electronics, agritech, and digital services. Instead of competing in isolation, African countries can specialize and connect, linking research capacity in South Africa to software development in Kenya, hardware assembly in Egypt, and digital finance in Nigeria. Such cross-border integration can reduce import dependency, lower production costs, and generate high-skilled employment.

To facilitate this, Africa must invest in interoperable infrastructure, data centers, regional broadband corridors, and harmonized digital identification systems to enable the free flow of goods, services, and data. The Smart Africa Alliance and the African Union's Digital Transformation Strategy (2020–2030) provide critical blueprints for achieving this connectivity. Policies that simplify cross-border intellectual property protection, digital taxation, and e-commerce regulation will further enhance investor confidence and entrepreneurial collaboration.

Regional innovation corridors such as the Eastern Africa Innovation Hub Network and the ECOWAS Centre for Renewable Energy and Energy Efficiency (ECREEE) demonstrate how shared platforms can promote collective technological advancement. Scaling these initiatives will help Africa transition from a consumer of imported innovations to a producer of integrated technological solutions that serve regional and global markets.

7.3 *Encouraging Ethical and Sustainable Technology Governance*

Africa's digital future must be grounded in ethics, inclusion, and sustainability. As technologies like artificial intelligence, biotechnology, and big data reshape daily life, the continent faces new moral and regulatory questions about privacy, bias, surveillance, and environmental impact. Without proactive governance, these technologies risk reproducing global inequities under the guise of innovation.

Building ethical technology systems requires the development of context-sensitive regulatory frameworks that reflect African values and realities. Governments should establish independent data protection authorities, promote algorithmic transparency, and enforce environmental standards in ICT manufacturing and e-waste management. Ethical AI principles such as fairness, accountability, and inclusivity must be embedded into national digital strategies and corporate governance codes. The African Union's Convention on Cyber Security and Personal Data Protection (Malabo Convention) and the AU Data Policy Framework provide vital policy anchors but must now be domesticated and enforced across member states.

Equally, sustainability must be central to technology adoption. Renewable energy should power digital infrastructure; circular economy principles should guide ICT production and recycling; and green innovation should be incentivized through tax benefits and carbon credits. Africa can position itself as a leader in sustainable digitalization, where technology not only drives growth but also mitigates climate risks and conserves resources. This alignment between innovation and sustainability will be critical for meeting both Agenda 2063 and the UN Sustainable Development Goals.

7.4 Strategies for Fostering Homegrown Innovation and Resilience

For Africa to achieve true technological sovereignty, it must nurture homegrown innovation ecosystems that draw on local knowledge, indigenous creativity, and regional collaboration. Building resilience begins with empowering African entrepreneurs to design solutions for African challenges, supported by policies that reduce dependency on external technologies.

Governments should prioritize the creation of innovation districts, science parks, and start-up cities that connect universities, industries, and investors. Rwanda's Kigali Innovation City, Egypt's Smart Village, and Nigeria's Eko Innovation Hub illustrate how spatial clustering can accelerate innovation spillovers. These hubs must be complemented by national procurement policies that give preference to local technologies and enterprises to stimulate domestic demand for innovation. Strengthening indigenous knowledge systems and integrating them with modern science can also fuel sustainable innovation. Traditional architecture, agriculture, and medicine offer proven, ecologically adapted solutions that can be digitized, scaled, and commercialized. Protecting intellectual property for community-based innovations through protocols such as Swakopmund ensures that local inventors share in the value they create.

Moreover, Africa must enhance its crisis resilience through digital sovereignty. The COVID-19 pandemic and recent geopolitical disruptions underscored the vulnerability of import-dependent economies. Investing in local manufacturing of ICT components, pharmaceuticals, and renewable energy systems will reduce exposure to external shocks. Likewise, building domestic capabilities in data analytics, cloud services, and cybersecurity will safeguard national interests and foster public trust in technology.

Above all, promoting homegrown innovation requires a cultural shift, one that celebrates experimentation, tolerates failure, and rewards creativity. Governments and media must reposition science and entrepreneurship as aspirational career paths. Educational systems should cultivate curiosity and entrepreneurial thinking from early stages, while financial systems must accommodate risk-taking through venture funds and innovation grants. When innovation becomes both a national priority and a social value, resilience follows naturally.

8 Conclusion

Africa's technological evolution tells a story of resilience, creativity, and transformation. It is a story that defies the narrative of perpetual dependency and instead reveals a continent steadily building the capacity to define its own digital destiny. From the indigenous innovation traditions that predated colonial rule to today's thriving start-up ecosystems and policy experiments, Africa's technological environment reflects both continuity and reinvention rooted in local realities yet increasingly global in scope.

This chapter has demonstrated that understanding Africa's technological trajectory requires viewing it through three interconnected lenses of history, policy, and innovation. Historically, Africa's relationship with technology was shaped by colonial extraction, postcolonial industrial ambitions, and the neoliberal reforms that followed. These legacies continue to influence how knowledge is produced, owned, and applied. The policy aspect shows uneven but notable progress: most African countries now have dedicated STI frameworks, regional cooperation mechanisms. However, as the analysis indicates, implementation remains the Achilles' heel.

In sum, Africa's technological story is still unfolding. It is not a narrative of belated participation but of emergent leadership, one that bridges ancient ingenuity with modern science, local challenges with global opportunity, and progress with humanity. The task ahead is clear: to sustain this momentum through inclusive governance, bold investment, and a shared continental vision that makes Africa not just a beneficiary of the digital age, but one of its architects.

References

African Continental Free Trade Area Secretariat. (2023). *Protocol on intellectual property rights*. AfCFTA Secretariat.

African Union Commission. (2015). *Agenda 2063: The Africa we want*. African Union Commission.

Freeman, C. (1987). *Technology policy and economic performance: Lessons from Japan*. Pinter.

Freeman, C. (1988). Japan: A new national innovation system?. In G. Dosi, C. Freeman, R. R. Nelson, G. Silverberg, & L. Soete (Eds.), *Technical change and economic theory* (pp. 330–348). Pinter.

Government of Kenya. (2007). *Kenya Vision 2030*. Government Printer.

Government of Nigeria. (2022). *National policy on science, technology and innovation (2022–2030)*. Federal Ministry of Innovation, Science and Technology.

Holl, A. F. C. (2009). *Early African metallurgy: The origins of iron smelting and working*. AltaMira Press.

International Energy Agency. (2022). *Africa energy outlook 2022*. IEA.

International Monetary Fund. (2003). *Trade and growth in Africa: Structural adjustment revisited*. IMF.

Lundvall, B.-Å. (Ed.). (1992). *National systems of innovation: Towards a theory of innovation and interactive learning*. Pinter.

National Planning Commission. (2012). *National development plan 2030: Our future – Make it work*. Government of South Africa.

Nelson, R. R. (Ed.). (1993). *National innovation systems: A comparative analysis*. Oxford University Press.

Phillipson, D. W. (2005). *African archaeology* (3rd ed.). Cambridge University Press.

Schmidt, P. R., & Childs, S. T. (1995). *Creativity, culture, and iron technology in Benin, Tanzania, and Uganda*. Indiana University Press.

United Nations Economic Commission for Africa. (2021). *Digital transformation strategy for Africa (2020–2030)*. UNECA.

Chapter 2
From AI Workflow Automation to Acceleration: The Impact of Artificial Intelligence on Sustainable Business Growth in Africa

Francis Okoye

Abstract Artificial Intelligence (AI) is transforming Africa's economic landscape by changing how companies automate workflows, allocate resources, and build sustainable businesses. This chapter examines how African companies have automated workflows with AI, beginning with digital process initiation and automation and advancing to the intelligent acceleration stage. Drawing on socio-technical systems theory, innovation diffusion theory, and dynamic capabilities theory, the author integrates conceptual and empirical evidence from the finance, agriculture, health, and logistics industries. The estimates show that automation improves productivity by 40% and lowers operational costs by 25%, while also increasing innovation and environmental effectiveness. However, significant barriers to AI adoption remain for many potential users, including infrastructure, human capital, and governance deficits. The research argues that proper human-centered, sustainable AI adoption is only possible if the political aspect is ethical governance, the innovation is inclusive, and the workforce is governed by reskilling. Moreover, it asserts that the use of automation is an imperative for the competitiveness of African enterprises in the global economy. The outlines of this chapter indicate that, in the absence of contrasting strategies, AI can help Africa grow at a sustainable rate while enhancing digital inclusivity.

Keywords Artificial intelligence · workflow automation · sustainable business growth · digital transformation · economic resilience

F. Okoye
Lagos Business School, Pan-Atlantic University
e-mail: fokoye@lbs.edu.ng

K. Modugu, J. Otalor (eds.), *Sustainability and Digital Transformation in Africa*, Frontiers in African Business Research,
https://doi.org/10.1007/978-981-95-9500-6_2

1 Introduction

Artificial intelligence is arguably one of the most disruptive new technologies of our age, as it augurs an unprecedented change in productivity and sustainability within all streams of the economy. AI is fast evolving on the continent of Africa, not only as an automator of desk-work but as an innovator in business intelligence and efficiency augmentation. AI is assisting in automating systems within the new business ecosystems on the continent, leapfrogging older and more inefficient systems in agriculture, healthcare, financial technology, and logistics (UNCTAD, 2023; World Bank, 2024). The focus of this chapter is on the use of artificial intelligence systems within the continent and whether these systems have helped African companies achieve sustainable growth, reduce inefficiencies, and improve global competitiveness. The chapter highlights the movement within the continent from more basic forms of robotic process automation (RPA) to sophisticated systems that demonstrate intelligence, learn, and make contextual decisions on their own (PwC, 2022).

Despite Africa's tremendous growth in technology, African businesses and the general economy are still faced with evolving challenges. According to the African Development Bank (2023), the most glaring factors are the form of governance, the digital infrastructure, and the technical skills available. Most businesses within this domain are still constrained by factors such as non-existent AI professionals, interrupted electrical supply, and exorbitantly priced internet access. Though the potential for AI to accelerate business transformation comes with the additional promise of sustainability. However, the fulfillment of this potential is contingent on the coherent formulation of strategies corresponding to the actual implementation of the technologies within the local geopolitical, policy, and ethical contexts (OECD, 2023). The aim of this chapter is to trace the evolution of the automation of AI workflows as a driver of business and sustainable development on the African continent. It presents a conceptually and empirically integrated account of the impact of AI systems on operational workflows, decision-making, and resilience within African businesses. It also focuses on the ethical, policy, and organizational elements essential for responsible and equitable AI adoption.

In addition to productivity and competitiveness, workflow automation through artificial intelligence is becoming one of the most important environmental management tools of African companies. The intelligent automation enables the firm to lower carbon emissions and material waste, as well as optimize the use of energy by substituting reactive processes that consume resources by using data-driven predictive decision-making. In the logistics, farming, production, and power-intensive services, AI-based route optimization, demand prognostication, and predictive maintenance minimize fuel consumption, unused capacity and excessive manufacturing. With African economies facing climate risks and an increase in the cost of energy, AI workflow automation offers an avenue to align business performance to green growth and Environmental, Social, and Governance (ESG) commitments, and rebrand sustainability as a factor of operation and strategic benefit.

2 Literature Review

2.1 Conceptual Review

2.1.1 Conceptualizing AI Workflow Automation

AI workflow automation depicts the application of artificial intelligence—specifically, machine learning (ML), natural language processing, and computer vision—to process management systems for data-driven, adaptive, and autonomous operations (Brynjolfsson & McAfee, 2021). While traditional automation is bound by fixed rules and deterministic logic, the application of workflow automation infused with AI incorporates cognitive features that enable systems to interpret, learn, and optimize (Davenport & Ronanki, 2022). AI-enabled automation in business enhances decision-making, predictive analytics, and customer engagement while minimizing human error and operational delays (PwC, 2022). For example, processing customer invoices, tracking supplies, and answering queries can be automated through AI's intelligent agents designed for natural language interaction and contextual engagement (McKinsey, 2023). The integration of AI and workflow management technology brings a new perspective on geo and business automation. Intelligence has replaced efficiency and is now the driving component of value. AI is now seen as a transformational pillar of business automation and strategy.

2.1.2 Sustainable Business Growth

Sustainable Business Growth (SBG) is the phenomenon where companies create more value economically and at the same time protect social and environmental well-being (OECD, 2023; UNCTAD, 2023). Unlike traditional growth models, which focus on profits in the short run, SBG embraces the triple bottom line (TBL) approach, which partitions economic growth, ecological conservation, and social growth and equity (Elkington, 1997; Carroll & Brown, 2018). This approach signifies how the purpose of a corporation has evolved from being strictly profit-focused to value co-creation for every single involved party. The concept of SBG is based on the fact that a company's long-term competitiveness is determined by its ability to responsibly innovate, dynamically adapt, and function within the boundaries of natural and social systems (Dyllick & Muff, 2016). Companies that strategically integrate the principles of sustainability into their major functions usually exhibit better resilience, risk management, and stakeholder trust (Porter & Kramer, 2019). Business growth is now viewed from a more strategic angle, in which SBG is seen as a multifaceted concept that transcends financial gain to include social and ecological well-being as major determinants of organizational success.

Innovation continues to be the primary driver of sustainability, allowing firms to redesign products, services, and processes to minimize negative environmental impacts while simultaneously capturing value (Hart & Dowell, 2011). Today, in the digital era, firms are able to advance sustainability due to the use of technologies such as artificial intelligence (AI), resource optimization (big data), and blockchains, which utilize data optimally to foster better decision-making (McKinsey, 2023; PwC, 2022). Hence, proper stakeholder management increases legitimacy and enhances corporate image. Through community participation, firms promote social trust by implementing the ethical employment of labor and inclusive supply chains (Freeman, Harrison & Zyglidopoulous, 2018). Environmental, Social, and Governance (ESG) corporate governance frameworks sustain the institutionalization of accountability and ethical integration by deepening the business strategy sustainability to foster oversight in accountability (OECD, 2023).

In Africa, the growth of the business sector is sustainably tethered to the developing challenges and opportunities the continent has to offer. The business in the region operate under highly dynamic set ups with a poor infrastructure, a multitude of informal sectors, and a continuing increase in the use of digital technologies (World Bank, 2024). Therefore, the efforts toward sustainability are targeted to, designed for and developed under the auspices of economic growth, social inclusion, and environmental protection. African firms are able to utilize a digital and green resource to promote sustainability through innovations. Fintech platforms such as Flutterwave and M-Pesa enhance affordability of payment systems which fosters financial inclusivity while agritech startups like Twiga Foods and Zenvus apply AI and data analytics to improve agricultural productivity and minimize waste (AfDB, 2023; UNCTAD, 2023). Similarly, M-KOPA and d.light renewable energy companies are leading the way in off-grid solar systems that improve energy access and decrease carbon emissions at the same time. Yet, the uneven enforcement of regulations, lack of funding for green innovations, and low levels of technical expertise are structural barriers that slow down progress (OECD, 2023). These barriers can best be tackled through comprehensive integrating and industrial policies that support the integration of sustainability, foster public–private collaborations, and encourage responsible private and public governance.

Sustainable business growth (SBG) can only be achieved through a combination of growth and equilibrium and requires diverse indicators. Along with profitability and market share, various financial metrics such as sustainability, climate, member and resource retention, carbon, and community impact of the organization are also used (UNCTAD, 2023). Many enterprises are starting to implement the Environmental, Social, and Governance (ESG) compliance reporting frameworks as advocated for by the Global Reporting Initiative (GRI) and the United Nations Principles for Responsible Investment (UNPRI) in a bid to enhance the integration of the sustainability social reporting. In an African context, sustainability also involves assessing the contributions toward the national and regional development goals, particularly the African Union Agenda 2063 and the United Nations Sustainable Development Goals (SDGs). These frameworks also set limits for the synchrony between business performance and social inclusion, innovation-driven environmental conservation, and growth. Growth in business sustenance promotes long-term competitiveness. This is achieved by an organization embedding adaptability, innovation, and legitimacy into its DNA. Organizations that embrace sustainability internalize it; thus, they show greater resilience to market volatility, changes in regulations, and environmental disruptions (Porter & Kramer, 2019). Furthermore, these firms also enjoy greater investment, talent retention, and customer loyalty which are essential in an increasing global economy that is digitizing rapidly (McKinsey, 2023). Within an African context, the intersection of sustenance and digital transformation offers an unparalleled leapfrogging opportunity, that is, transcending the costly process of industrialization and embracing inclusive, green, and technology-intensive development. Enterprises in Africa are able to achieve scalable impact by integrating AI and automation into their sustainability strategies which also accords with global frameworks for responsible innovation.

Innovation is the main focus of sustainability that allows the firms to re-architecture products, services and workflows within their operations to minimize environmental footprints and gain economic value (Hart & Dowell, 2011). Artificial intelligence can be used to further this ability in the digital age having environmental intelligence built directly into workflow automation systems. Analytics through AI will allow companies to track real-time energy usage and use inputs optimally, predict waste production and cut down on emissions along value chains. Investigating, demand forecasting will decrease overstocking and end-of-harvest waste, whereas predictive maintenance will decrease the energy-intensive breakdowns and timely replacement of equipment. These processes enable AI workflow automation to shift sustainability as an external reporting practice to an internal operation rationale, enhancing the internalization of the ESG principles in daily business operations (OECD, 2023; McKinsey, 2023).

2.2 Theoretical Foundations

The Socio-Technical Systems Theory holds that organizational success is contingent upon the interplay of the technological framework and the human processes (Trist & Bamforth, 1951; Pasmore, 2015). In the African context, AI and human skills, cultural frameworks, and governance systems balance the value of technology integration, especially AI, as seamless integration of technology in an organization is governed by the level of social sophistication such an organization is able to achieve. Supporting this view, Rogers' (2003) Innovation Diffusion Theory describes the channels by which innovations are disseminated, adopted, and embedded in social systems. This lens is useful in assessing the context in which African countries are embracing AI automation with varying infrastructural developments, levels of awareness, and attitudes of appreciation. It highlights on technology adoption that the social context, the perceived value of the technology, and the level of technological integration with other systems are critical. Dynamic Capabilities Theory (Teece, 2018) concludes this theoretical framework by underscoring the integrated strategic ability of a firm to incorporate, build, and reconfigure internal and external resources to rapidly dynamic demands in the environment. In this regard, AI automation acts as a driver of augmented dynamic capabilities by facilitating evidence-based decision-making, innovation, and perpetual organizational evolution. Altogether, these theoretical perspectives offer an integrated base for analyzing the relations among technology, human action, and organizational context for shaping AI-driven change. They all explain how automation of workflows stimulates organizational learning, the effectiveness of business operations, and sustainability in developing economies in Africa.

2.3 *AI and Sustainable Business Growth in Africa*

The implementation of artificial intelligence in Africa stems from the continent's digital transformation framework which aims to use technology for economic advancement (UNCTAD, 2023). When compared to other regions in the world, the digital economy makes a mere 5% contribution to the continent's GDP, but AI is expected to inject over $1.2 trillion into the continent's economy by the year 2030 (World Bank, 2024). AI is used to enhance the value chain in agriculture, logistics, and even fintech. AI also increases the availability of operational and financial intelligence. Take, for example, Flutterwave and Paystack which are fintech companies and AI's use in fraud detection and improving customer increase services and satisfaction (McKinsey, 2023). Kobo360 is another logistics company that uses AI to proactively predict and analyze optimal delivery routes. Furthermore, there are AI agriculture platforms like Zenvus which use soil and climatic data to make recommendations that help to maximally produce food in a sustainable manner.

The world is still facing many of social and economic inequalities. Geographical inequalities in the distribution of digital and AI technology is a major discrepancy. Fewer than 40% of African firms have access to fully-fledged automated systems (WEF, 2023). Beyond this, the disparity that exists between rural and urban regions is a source of concern as it threatens to worsen inequity, a major hindrance to realizing the potential of AI. Therefore, there should be significant social and policy changes to ensure equity and better AI adoption.

2.4 *Ethical and Governance Considerations*

The use of AI technologies in Africa comes with a tangled array of ethical and regulatory challenges concerning privacy, biases in algorithms, and questions of responsibility. Studies suggest that AI systems which are developed using poorly representative datasets are likely to exacerbate already-present societal inequities (Eubanks, 2018; Crawford, 2021). With a predominately informal economy, Africa's poorly structured data environment increases the probability of producing erroneous outputs (OECD, 2023). Governance within the African Union, like the AI for Africa Strategy, articulates the necessity for AI policies which are open, comprehensive, and sensitive to local sociocultural contexts (AU, 2022). Ethics in AI systems demand that policies and frameworks are developed that incorporate local values, uphold data sovereignty, and equity in adjudication. It follows that the rapid progression of business via the use of automated AI systems depends on the alignment of legislation and active partnership among stakeholders from the public and private sectors.

3 Methodology

This chapter adopts an interpretivist approach to data, thus focusing on understanding rather than quantifying (Creswell & Poth, 2018). In order to increase the transparency and methodological rigor, this chapter uses the systematic qualitative synthesis and content analysis of the secondary data with the help of purposive case selection and triangulation. African enterprises were also gathered based on strictly outlined criteria: they must have adopted an AI-enabled workflow automation beyond simple digitization, must be relevant to major development sectors (finance, agriculture, logistics, and health), must be operated at a scale or have a scale impact, and must have secondary documentation that is credible. Based on this, examples like Flutterwave, Wasoko, Twiga Foods, Zipline, and Zenvus were used as examples of analytical representation to show the various phases of AI-powered automation and not as statistically representative companies. The literature of the study had reviewed more than 24 scholarly, policy, industry, and company documents published in the period of 2018–2025 and screened in accordance with their relevance, methodological soundness, and explicit connections to the productivity, cost-efficiency, sustainability, or governance outcomes. Data was cross-examined and cross-sectoralized to raise awareness and reduce the effect of single-source bias. Even though the research does not imply any statistical generalization, the transparent sampling rationale and systematic content analysis allow achieving a sufficient analytical rigor and theoretical strength to produce contextualized information about the challenge of AI-based workflow automation, as well as sustainable business development in Africa.

The content analysis approach to data analysis facilitated the examination of the data through the systematic identification of patterns and insights. The analysis concentrated on three core areas. The first examined the extent and the nature of the use of AI technologies in business operations in different sectors and how these technologies have transformed the processes of business operations. The second dimension looked at the correlation of AI-based automation with business sustainability and the impact of intelligent automation on productivity, cost, environmental responsibility, and resilience of organizations over the long term. The third dimension considered the ethical and governance issues on adoption, focusing on problems of data privacy, algorithmic bias, and business ecosystems in Africa concerning governance issues. The analysis of findings across these dimensions allowed the chapter to formulate a conceptual framework that associates the automation of AI workflows with the acceleration of sustainable business in Africa. This framework integrates the impact of technological adoption, ethical governance, and sustainability of framed AI-driven business processes on inclusive and resilient development of the emerging African economies. Reliability was achieved through the cross-checking of various data points and their consistency over time (Yin, 2018). For instance, the findings were in line with the established theoretical constructs of dynamic capabilities and the alignment of socio-technical systems in governance and boundary-spanning articulation. This approach, while lacking regional empirical datasets, ensures rigor and authenticity of the context and comprehensive qualitative research.

4 Findings and Discussion

4.1 *The Evolution of AI Workflow Automation in African Enterprises*

The implementation of AI-driven workflow automation in Africa was not an instantaneous development. There has been a decade-long development of technology and organization that can be described in three overlapping phases: digital initiation (from 2010 to 2018), process automation (from 2018 to 2022), and intelligent acceleration (from 2022 to current date). Each of these phases has distinctive technologies, patterns of adoption, enabling institutions, and results. Grasping this evolution explains why some sectors and firms were able to "leapfrog" legacy constraints while others continue to lag.

Figure 2.1 shows that the three-step model highlights that the process of environmental sustainability and ESG integration becomes most intense at advanced levels of AI maturity, which supports the strategic significance of intelligent acceleration to the green development path of Africa.

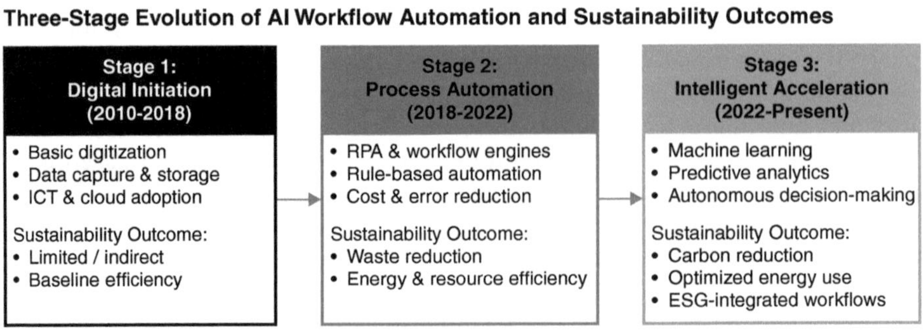

Fig. 2.1 Three-Stage Evolution of AI Workflow Automation and Sustainability

4.1.1 Digital Initiation (2010–2018)

In the initial stage, African companies implemented basic tools of ICT, which included mobile interfaces, emails, databases, and other useful record-keeping tools. This phase established the basic infrastructure and the culture required for further automation. This phase had two characteristics. One, the increase in mobile telecommunications and the availability of cheap smartphones developed a mass market for digital services, which made them commercially sustainable. Two, small and medium enterprises were able to subscribe to digital tools due to the cloud services, which eliminated the need for huge capital investments to set up in-house systems and services. Even though most tools implemented at this stage were limited to record management and communication, they were essential in laying the foundation for automation efforts.

4.1.2 Process Automation (2018–2022)

At this stage, there was a significant increase in the adoption of Robotic Process Automation (RPA) and workflow engine systems based on automation in the areas of high-volume repetitive tasks that offered the best return on investment on automation: the banking, telecommunications, large-scale retail, and distribution sectors were the earliest adopters. In Finance, banks automated customer service routing, KYC, and transaction reconciliation; in retail and logistics, platforms automated basic route scheduling, order processing, and invoicing (PwC, 2022). Enterprises benefited from RPA and workflow engine systems through reduced processing errors, greater turnaround time, and decreased back-office operational costs. However, the rigid automation systems that dominated this phase still required elaborate instruction sets. There was no escaping the use of manual work in case of rule-based decision-making systems, which still lacked sufficient intelligence to process exceptions.

4.1.3 Intelligent Acceleration (2022–Present)

Beginning in 2022, African companies started incorporating cognitive capabilities—such as machine learning, predictive analytics, natural language processing, computer vision, and predictive analytics—into workflow automation. The shift from automation as the execution of predefined rules to automation as providing decision support, or decision-making, describes what is known as "intelligent acceleration." Systems no longer perform rote tasks, but also learn from data, flag and bookmark anomalies, and recommend or perform appropriate, situationally aware, proactive actions. Faster scaling, more resilient operations, and the automation of workflows that were previously too unstructured or complex for RPA can be captured in the net outcome (McKinsey, 2023).

Notably, the smartening acceleration phase goes further than an increase in productivity to quantifiable environmental results. Workflow systems that use machine learning will enable companies to optimize the logistics routes, balance the loads of energy, and minimize idle operations, which will result in decreased fuel consumption and carbon emissions. AI-based load optimization and routing in logistics and distribution reduces unneeded trips and vehicle congestion, which directly reduces the intensity of emissions. In manufacturing and agriculture, AI processes reduce waste of inputs, excessive use of water, and spoilage. These sustainability benefits are not fleeting but systemic because smart automation is constantly informed by operational data to make environmental efficiency improvements over the years, which solidifies the connection between operational excellence and green sustainability objectives.

Based on the institutionally validated estimates, digital technologies accounted for around 5% of Africa's GDP, as assessed by the most recent analyses done by the World Bank, with AI and automation being the most crucial for further digital value creation (World Bank, 2024). The same literature cited regional forecasts that if the enabling conditions (connectivity, skills, governance) are met, AI-driven productivity increases could alone add USD 1.0–1.2 trillion to the continent's economy by 2030 (World Bank, 2024; McKinsey, 2023). In terms of enterprise readiness, most global assessments suggest that in the early 2020s, less than one-third of African firms had the requisite digital infrastructure and automation maturity to deploy advanced AI systems (WEF, 2023). This suggests both rapid progress and persistent gaps.

4.1.4 Case Study Examples

Financial services—Fraud detection and Flutterwave: Most large payment processors and fintechs on the African continent have incorporated AI systems for anomaly detection and real-time fraud scoring. Many Flutterwave customers have reported reduced chargeback rates, increased confidence in digital payments, and higher electronic commerce volumes owing to the reduced fraud (World Bank, 2024).

B2B distribution and logistics—Wasoko: Transitioning from scheduled logistics to dynamic, machine learning driven inventory and routing optimization is exemplified by Wasoko (formerly Sokowatch). Sales and delivery telemetry in conjunction with demand-forecasting algorithms have enabled Wasoko and other platforms to substantially lower out-of-stock events and delivery lead times. Diverse sectoral analyses suggest delivery time improvements of 20%–30% or more with machine learning driving real-time replenishment (McKinsey, 2023).

Healthcare logistics—Zipline: Zipline's drone delivery system in Rwanda and Ghana, while not strictly an "automation" in the narrow back-office sense, is an integration of routing algorithms, real-time telemetry, and automated dispatch, enabling faster and more reliable delivery of medical supplies to remote clinics; field evaluations reported significant declines in stockout and delivery lead times (World Bank, 2024).

Agritech and precision farming—Twiga and Zenvus: Agritech companies like Twiga and Zenvus now integrate machine learning models with routers, satellites, and IoT sensors to offer planting and irrigation assistance, pest forecasting, and market linkages. These services improve output and reduce losses. Evidence from pilot studies and platform reports shows post-harvest losses and yield enhancements, although the data varies by crop and deployment intensity (UNCTAD, 2023).

4.1.5 Enabling Factors and Ecosystem Dynamics

There are multiple reasons for the expansion of intelligent acceleration: first is the decline in expenses towards cloud computing and storage; the advent of mobile payments systems generating substantial volume of data for analysis for credit scoring and fraud modeling; second is the abundance of overseas AI tools and local adaptors and accelerators for AI startups; and third is the public and private partnership funding devoted to data and skills infrastructure (AU, 2022; AfDB, 2023). These factors have triggered a surge in entrepreneurship, with firms, especially in fintech, logistics, agritech, and healthtech being the first to showcase the returns on investments AI has brought to the continent on a large scale.

4.1.6 Persistent Constraints and Uneven Diffusion

Even with much-cherished advancements, the gap between digital "haves" and "have-nots" remains. Structural barriers—absence of broadband in remote regions, unpredictable power supply, exorbitant data costs, dearth of trained personnel in the AI field—continue to hinder the speed and outreach of AI adoption (UNESCO, 2023; AfDB, 2023). Legislative ambiguity (data protection, cross-border data flow, and algorithmic bias with insufficient or poor quality training data), and the primary investment to construct a model still limit the adoption of AI in MSMEs. As a result, the gap between "showcase" AH and "many modest" SBs is still too substantial.

The three-phase trajectory illustrates a critical lesson—the impact of AI is as much organizational and institutional as it is engineering. The "digital" step of the evolution architecture designs the data and connectivity foundation, the "process" step of the automation stage improves efficiency and builds confidence in automation, while the "intelligent" step of the acceleration stage facilitates value through prediction, adaptation, and scaling. For Africa, this means there is need for intertwined investments to expand connectivity and sustained power supply, build robust data governance and data skills, and cross-sectional projects with demonstrable socioeconomic value. It is with such systems in place that companies like Flutterwave, Wasoko, Zipline, and Twiga have been able to transform AI workflow automation to eliminate operational bottlenecks into absorbable and sustainable competitive edge. In the absence of any conditions, the diffusion will be shallow in nature, and the benefits of any outcome from the diffusion will be limited.

4.2 Drivers of AI Adoption

The growing dependence on artificial intelligence (AI) technology in Africa is attributed to high demands of operational efficiency, automation, cost constraints, and competition in both domestic and international markets. These phenomena corroborate the economic rationale of African firms' automation while at the same time, recognizing the paradigm shifts they undertake to deepen their agility, resilience, and datacentrism. As firms try to unlock bygone impediments of poor infrastructure, high transaction costs, and unproductive operations, workflow automation via AI has emerged as a prominent spearhead of productivity and innovative advancement.

4.2.1 Efficiency Gains

The desire for operational efficiency stands out to be one of the primary motivators for using AI in Africa. Many African businesses tend to function in conditions of extreme resource constraints and supply-chain fragmentation alongside "chains of command" that have historically suppressed productivity (OECD, 2023). Over and above, the efficiency of manual work that requires tactical strategic thinking and creativity is AI's forte. Routine, time consuming, and data driven work that can be performed in parallel to human labor, AI augments productivity. Automation of Advance processing systems in finance, logistic, and manufacturing works as an epitome of AI. For example, Equity Bank and KCB Group in Kenya harness the power AI tools to automate approvals and fraud detection within minutes (World Bank, 2024). In Guaranty Trust Holding Company (GTCO) in Nigeria, AI algorithms monitors and tracks transactions to aut identify compliance and human oversight errors (KPMG, 2022). This is to illustrate the gradual transition that many businesses are taking from being human-dependent to data-driven decision-making models.

The impact of efficiency gains is reflected greatly in operations that interface with customers. Chatbots and virtual assistants like Access Bank's TAMADA and UBA's Leo use natural language processing to respond to millions of inquiries, minimizing call center effort and response time by over 60% (KPMG, 2022; McKinsey, 2023). This reduces costs, improves service, and, along with the 60% reduction in response time, improves service. Employees now have the time to focus on the more sophisticated, relational, and analytical tasks. Moreover, in logistics and retail, AI allows for real-time monitoring of inventory and predictive demand forecasting. As a result, platforms like Wasoko and Kobo360 use predictive analytics and optimization to enhance vehicle delivery routing and usage, improving delivery time, fuel use, and customer satisfaction (McKinsey, 2023). There is a fundamental transformation in operations driven by AI, moving from manual and reactive to predictive and automated self-improvement. The gains in efficiency are concrete—process optimization is often in the range of 30% to 40%—and result in diagnostic value for large corporations as well as micro, small and medium enterprises (MSMEs).

4.2.2 Cost Reduction

AI automation is the second major driver after the adoption of the technology. It is also the most relevant element in Africa's highly competitive and resource-scarce business settings. Increased financial AI automation sustainability is derived from the lowering of mistakes, human labor, and resource allocation and the bettering of quality control (PwC, 2022). Fintech is a clear illustration. Back-office operations, like reconciliation, reporting, and customer verification, take a lot of time and resources. Access Bank, First Bank of Nigeria, and Stanbic IBTC have, for instance, improved compliance and accuracy while reducing operating costs performing RPA and AI driven data analysis (KPMG, 2022).

PwC's (2022) Africa Automation Report states that organizations that implement AI workflow automation are able to reduce costs of administrative and operational functions by 20% to 35%. Consider the AI invoice processors. These systems extract and validate financial data, anomalous and unmatched entries, and reconciliation with little to no oversight. In the manufacturing, energy, and telecommunications sectors, AI driven predictive maintenance also enhances cost efficiency.

For instance, South African mining companies employ machine learning to predict equipment failure and schedule maintenance before breakdowns occur, saving millions in downtime and repairs (WEF, 2023). MTN Nigeria and other telecom operators in Nigeria utilize AI-driven analytics to assess network performance and redistribute network resources, consequently decreasing operational costs and enhancing service (World Bank, 2024).

The advantages of cloud computing and Software-as-a-Service (SaaS) platforms further amplify cost benefits by minimizing capital outlays. Businesses can now access AI solutions such as virtual customer support, data analytics dashboards, and automated accounting systems. Scalable, sophisticated automation tools are available via subscription-based models, making them affordable for even nascent companies and MSMEs. Ultimately, AI's potential for cost reduction enables African enterprises to remain competitive, regardless of changes in the macro economy, currency volatility, and high import prices. Integration of AI into workflow systems enhances productivity and resource optimization, resulting in competitive favorable taxation.

4.2.3 Market Competitiveness

The third driver for AI adoption is the quest for competitive markets, which pertains to the capacity of companies to distinguish themselves through higher levels of intelligence, versatility, and innovativeness. Firms that incorporate AI in the globalized digital economy possess a substantial advantage in understanding consumer behavior, anticipating demand, and providing tailored products and services. Firms' overall competitiveness is enhanced through AI driven decision-making. In addition, predictive analytics and machine learning enables firms to identify nascent market dynamics and trends, effectively segment customers, and refine marketing initiatives. As an illustration, Jumia, the largest e-commerce platform in Africa, utilizes AI in product recommendation systems and personalizes user engagement to predict user actions and enhance conversion and retention rates (UNCTAD, 2023).

In the telecommunications industry, predictive customer churn management and individualized engagement systems. AI tools employed by Airtel Africa and MTN Group accomplish as much as 15% churn reduction in retention incentive markets through machine learning that identifies customer dissatisfaction and customizes retention offers (McKinsey, 2023). AI's untapped potential in agriculture and supply chains through real time data is very crucial for enhancement of competitiveness. AI assists with logistics, computing the demand for crops, pricing, and integrating data which helps the farmers and distributors make profitable decisions with minimal delay, and therefore, helps in increasing profits and avoiding wastage (World Bank, 2024; UNCTAD, 2023).

Furthermore, AI enables African Countries to integrate into the global value chains (GVC) with the required quality, transparency, and sustainability needed for global value chains. More and more African exporters are installing automated compliance systems, predictive quality assurance, and carbon-footprint monitoring to better meet the global ESG (environment, social, and governance) standards (OECD, 2023). All in all, AI serves as a source of trade competitive advantage which enables firms to predict changes, deploy quickly to changes, and advance on the market to bring new things, which are necessary thoughts in the rising market.

4.2.4 Scalability and MSME Empowerment

Beyond the boundaries of efficiency, cost, and how competitive the business is, scalability is an important driver of AI adoption in Africa, especially for MSMEs that constitute over 90% of the continent's enterprises and employ nearly 60% of its workforce (AfDB, 2023). MSMEs have been limited in growth in the past because of the growth in the scale and the linear relationship between labor—the scale of operations went up, so the number of employees, the infrastructure, and the costs needed to make operations run had to grow in a set ratio. AI changes this scenario by providing non-linear scalability, which allows for the growth of output dramatically without associated increment in input. Cloud-based AI tools of a low cost available to MSMEs which enable the automation of a range of useful functions include: Zoho CRM, QuickBooks AI, and Microsoft Azure Cognitive Services. They enable the automation of the tracking of inventories, customer relations, and performance of Financial Analytics. A case in point, AI powered point of sale systems deployed by small retails in Kenya enable sale data being accumulated to be auto updated to cloud dashboards, the dashboards in turn provide analytical insights which in the past needed a lot of human input and supervisors.

Also, the use of AI is an alternative way of democratizing access to financial services, particularly through the use of alternative credit scoring systems. In Nigeria, the AI model used by the two hundred and thirty-four funded startups including Migo and Carbon analyze mobile and transaction data for prospective borrowers, thereby filling the gap left by traditional finance systems for small and underserved entrepreneurs (World Bank, 2024). AI adoption enhances access to finance, thus reinforcing the resilience and inclusion of MSMEs. It also helps underserved entrepreneurs access the needed financing in a range of economical and responsive ways. The AI adoption in Africa for efficiency, cost, and competitiveness and scalability also allow for non-linear growth. The scale, however, is limited by both strategic and practical reasons. The reasons behind the adoption of AI by enterprises go far beyond simply automating processes. There is operational DNA transformation to be considered. Benefits derived from operational efficiency and cost reduction are immediate and tangible. In the same breath, long-term competitiveness and market relevance porter's five forces model drives the need to sustain market relevance. Scaling the stated drivers of transformation is worrisome from the policy and capability gaps. Huge portions of enterprise infrastructure, skilled personnel, and governance frameworks to sustain intelligent automation at scale are deficits. These deficiencies need to be addressed through investment in digital infrastructure, data governance, and AI education in order to maintain momentum. AI adoption within African enterprises is not a periphery adoption trend. It is structural evolution toward a business model that is smarter, adaptive, and more globally competitive. African enterprises have taken the bold step of positioning themselves at the frontier of the next industrial transformation. This transformation is the result of the intelligent inclusion and sustainable growth.

4.3 Barriers to AI Integration

While these benefits are salient, the adoption of AI within Africa still remains inequality distributed due to issues arising within infrastructure, finances, and human capital. The World Economic Forum (2023) states that only 28% of businesses on the continent have the digital infrastructure necessary to implement advanced automation. Limited broadband access and unstable, and expensive electricity, and high data costs still constrain the use of AI in African businesses (World Bank, 2024). Human capital presents an even greater challenge, as Africa faces a significant shortage of AI engineers, data scientists, and cybersecurity specialists (UNESCO, 2023). Although the continent lags behind Asia and Europe in the development of AI capabilities (African Union, 2022), initiatives such as Google's AI research center in Ghana and the African Masters in Machine Intelligence (AMMI) are playing a crucial role in building local expertise and strengthening the region's AI capacity. There is also a lack of examples on legal and ethical AI use. Most African countries have yet to implement comprehensive AI governance policies, and as such, enterprises are exposed to the risk of data misuse, cybersecurity breaches, algorithmic bias, and other risks (OECD, 2023). There are fears that the absence of stringent data protection policies in place, the adoption of AI is likely to heighten digital inequality.

4.4 Sectoral Insights

The adoption of AI in workflow automation has advanced business growth and operational agility in Africa's key sectors—banking and finance, agriculture, healthcare, and logistics. AI's impact has been quite significant. The use of AI technology in logistics allows optimizing the performance of operations, both in terms of cost and time, and in terms of environmental performance. Kobo360 is an example of a platform that uses AI-based demand projections, route planning, and fleet tracking to minimize empty returns, fuel usage, and vehicle unproductive working time (McKinsey, 2023). These systems help reduce the amount of carbon emissions per delivery by reducing inefficient routing and unnecessary mileage in addition to enhancing asset utilization. Since the logistics sector continues to be among the most rapidly increasing contributors to emissions in Africa, AI-based workflow automation provides a decentralized system of scaling economic growth–environmental degradation decoupling.

AI has proven indispensable in augmenting fraud detection, user profiling, and assessing credit risks. Real-time transaction authentication and fraud detection systems used by Flutterwave, Paystack, and other fintech companies rest on advanced machine learning (KPMG, 2022). Digital finance transactions security, as well as consumer confidence needing critical restoration, has also been boosted by such innovations, a trend much desired in Africa's expanding digital economy. As transaction volumes increase, institutions gain operational resilience through cost and compliance automation, which leads to improved transaction accuracy. This also promotes financial inclusion and economic resilience of the region.

AI-powered platforms like Twiga Foods and Zenvus are transforming food and agriculture supply chains in Kenya and Nigeria through predictive analytics and smart farming. Through the application of the Internet of Things (IoT) technology, coupled with machine learning, farmers are now able to assess the quality of their soil, yield forecasts, and even automate the irrigation and fertilization of their crops (UNCTAD, 2023). This, in turn, satisfies the disutility maximization while enhancing the utilization of the available resources. The approach also aims to achieve the first and third Sustainable Development Goals (SDGs) of the United Nations, which are zero hunger and responsible consumption, respectively. Consequently, automation of AI techniques tackles an even greater problem of food security, along with improving the sustainability of agriculture on the continent. In addition to productivity improvement, the direct benefits of these AI-powered agricultural processes are related to environmental sustainability in that post-harvest losses are minimized, fertilizer and water usage is optimized, and food waste and inefficient transportation results are reduced. The agritech platforms have shown the way AI workflow automation can deliver food security and climate resilience goals in African economies by integrating sustainability metrics into their operational decisions.

AI automation has also shown a significant positive impact on the healthcare industry, particularly in terms of diagnostics, patient management, and the provision of other health services. Healthcare practitioners are supported by AI in the analysis of medical images to find and flag certain anomalies, as well as in the triage of patients so as to provide timely care. One of the prominent cases is the Rwandan company Zipline, which uses drones to deliver medical supplies to underserved communities and relies on AI to optimize its logistics networks. Integration of AI with drones has also enabled up to 70% reduction in delivery times, thus enhancing access to essential medicines, blood, and vaccines in the most remote and rural areas (World Bank, 2024). This is just one instance of how automated systems AI has contributed to improving public health coverage, which translates to equity in health access.

AI technology in logistics, similarly, enables optimized operations and networks, as well as demand forecasting and the reduction of operational inefficiencies. Kobo360, for example, employs AI for demand forecasting, route optimization, and real-time supply chain visibility analytics systems to enhance operational productivity for cross-border freight services (McKinsey, 2023). These systems evaluate complex data sets, including traffic patterns and fuel consumption, and analyze delivery and performance data to prescribe cost-effective routes and resource use. These changes also improve time, cost, and energy efficiency and, in turn, profitability and competitiveness, which decreases carbon emissions from logistics operations.

Overall, these use cases illustrate that the economic gains from the use of AI that are frequently highlighted are only part of the story. Such innovations also enhance sustainability through reductions in energy and waste, as well as greater participation in the economy. For African companies, AI reduces the manual effort required to devise systems that are "intelligent" so that the resultant business models are adequately innovative for competitiveness, as well as socially and ecologically responsible. Thus, AI is both a driver of economic change and a sustainable development accelerator for the continent.

4.5 Ethical and Societal Implications

AI combines unprecedented opportunities for innovation and sustainable growth while offering deep ethical and societal challenges that need a careful policy framework and institutional supervision. The swift emergence of AI automated workflow across Africa raises the issues of job loss, algorithm bias, lack of data control, privacy, and blurred accountability. Addressing these challenges is fundamental for steering AI in Africa towards inclusive growth and development, as opposed to deepening inequalities.

The loss of jobs, together with the future of work, is one of the most critical challenges. Recent research by McKinsey predicts the candidates shortlisted for 2030 are poised to automate as many as 20 million jobs in the manufacturing and administrative sectors. The scenario is particularly troubling for Africa, whose average rates of joblessness stand at more than 7%, alongside an exceedingly high youth jobless population. This phenomenon is known as the ILO, 2022. It is equally important to note, however, that the consequence of automation is not merely the reduction of the total number of jobs, but the redefinition of the nature of given jobs. AI, for example, replaces numerous manual jobs with more advanced positions in digital technology, data engineering, and machine control supervision. These positions necessitate advanced cognitive and technical skills. To avoid the social dangers posed by automation in Africa, most of them, if not all, the government and the private sector are urged to embrace AI with the framework of the "human-centered approach."

The direct incorporation of digital literacy and the ability to adjust and transform toward future technologies as part of the AI National policy and digital economy framework of Nigeria demonstrates the prioritization of these skills. "AI for Africa" and Microsoft's Africa Development Center, and similar cross-border projects, target skills gaps in younger and older generations in coding and AI to empower them in the digital economy. This targeted advancement of human capital serves as an example of how socioeconomic disruption caused by rapid technological shift can be transformed into an advantage.

Cascio and Boudreau (2019) describe the computation of basic data of the workplace as an issue of great ethical concern. The shift concerning algorithm bias as one of the topics of concern is indeed of great importance. African countries and populations are often the subjects of studies that suffer from neglected datasets, which render theories and models developed to them inaccurate and even harmful. The socio-technical system of injustices and inequities caused by technological advancement is more amplified with the harsh reality of economic disparities in the rest of the world. There is also the example of AI and the case of discrimination in the use of face recognition systems, which are argued to be unfair and unreliable. Foundations of social injustices are often ignored in the discourse of economic disparities. Countless people are subjected to harsh discrimination by the use of the so-called "intelligent" systems. For example, credit-scoring systems in the financial services industry, managed by a dominant paradigm and a lack of sufficient data, impose restrictions that are unfair.

This challenge stems from the decentralization of data, as well as the creation of contextually relevant AI models, which encompass the significant diversity of Africa's social and cultural landscape. Also associated with algorithmic bias is the problem of data sovereignty and privacy. There is the concern of who owns, controls, and has the right to use the data collected in Africa, considering the AI systems in use rely heavily on data collection and processing. There are many African countries that are still reliant on cloud services far outside the continent, putting them in a position of dependency, which threatens their national security and economic sovereignty. Data Protection, Personal Data, and Data Governance in Africa (2022) and the Malabo Contract on Cyber Security and Data Protection (2014) from the African Union still stress the need for the ethical and human rights-respecting governance of data, to contain personal information, and to promote the establishment of local data centers. To make sure that the use of AI systems is beneficial to the country and does not undermine the country's sovereignty, there is a need to build national and regional infrastructural data protection and trust concerning AI systems.

Applying ethical AI includes the obligation to promote and implement design and system use that has social justice, multi-stakeholder, and multi-disciplinary concerns. The processes of AI decision-making can be understood and examined through the lens of transparency, while accountability covers the assigning of responsibility in the event of errors or harm caused by automated systems. Inclusivity, on the other hand, ensures that the implementation and governance of AI technologies incorporate a range of stakeholders, particularly those from marginalized communities (OECD, 2023). Therefore, a coalition among governments, academic institutions, and civil society is required to construct ethical governance systems that integrate artificial intelligence alongside social justice and the dignity of all human life.

The AI for good paradigm, supported by the UN and other development partners, captures the ethical obligation of crafting artificial intelligence systems that promote equity and empowerment. It involves focusing on the most critical areas in Africa, including but not limited to the social issues of education, healthcare, the economy, and climate change, within a human rights and social cohesion framework. It is only by ensuring ethical AI governance across the entire continuum of adoption—from data collection and algorithm design to implementation on the continent—that Africa will be able to harness artificial intelligence for development.

AI workflow automation is, undoubtedly, transforming the business landscape of Africa. However, the ethical and social implications of artificial intelligence should not be neglected. Without effective governance in place, artificial intelligence will serve only a small section of society, increasing inequality and a general disillusionment with technology. With proper governance, equitable access to digital education, and localized data governance, Artificial Intelligence can propel progress that is human-centered, community-empowering, right-protecting, and economically sustainable across the continent on the condition that governance remains untainted and data education access unrestricted.

4.6 The Strategic Imperative of AI for Sustainable Growth

In Africa, artificial intelligence (AI) is used as a tool to facilitate sustainable growth across various sectors due to its ability to improve productivity, optimize resource utilization, and foster adaptive capacity. Harnessed properly, the workflow automation offered by AI is geared towards promoting economic resilience, environmental preservation, and innovation. These have been identified as the three pillars on which sustainable growth rests. Automation driven by AI strengthens economic resilience by increasing the ability of firms to weather market fluctuations and external shocks. For centuries, Africa has suffered the brunt of geopolitical disturbances and the trickle-down effects of infrastructure deficits and shifts in commodity prices. AI equips companies to develop flexible and data-driven adaptive systems that mitigate risk. For instance, predictive analytics assists firms in foreseeing demand, disruption in the supply chain, and in adjusting production schedules. This kind of agility was largely absent during the COVID-19 pandemic, especially in Africa, where companies that advanced in digitization and AI were able to operate remotely and quickly reorganize supply chains compared to their counterparts (World Bank, 2024). The ability to move from reactive to predictive and adaptive business models augers well for AI-driven companies in operational continuity, strategic foresight, and competitive advantage.

AI adoption is also associated with improved productivity and economic performance, which is demonstrated by the increased efficiency gains that have been recorded. As shared by PwC (2022), businesses that use AI workflow automation systems describe improvements of 40% in efficiency and a 25% decrease in operational costs. These improvements stem from the automation of repetitive jobs, the better use of resources, and the decrease in waste. For instance, in South Africa, manufacturing firms use AI-driven systems that conduct real-time predictive maintenance of machinery, thereby improving asset availability and life cycle due to reduced downtime. Likewise, logistics firms like Kobo360 apply AI to optimize the matching of cargo with transporters, reducing fuel costs from nonproductive trips. These improvements in efficiency not only increase profits but also help in achieving sustainability goals by saving energy and reducing the negative impact on the environment.

AI also represents the pillar of sustainability. In all sectors, companies are intensively using AI to optimize the use of energy, control emissions, and manage resources. AI enabled systems help firms to manage energy by analyzing use patterns, forecasting peak consumption, and wasteful operational intervals. In Kenya and Nigeria, smart-grid technologies apply machine learning techniques to balance the national electric grids and more effectively integrate renewable energy resources (UNCTAD, 2023). Moreover, AI-assisted carbon tracking systems enable corporations to align the reduction of greenhouse gas emission targets with the international climate objectives, the Paris Agreement to the UN Sustainable Development Goals.

The automation of AI workflows also empowers the integration of ESG since it allows companies to monitor, confirm, and transmit the indicators of environmental performance in a systematic way. The use of automated data capture and analytics enables organizations to track the emissions, energy consumption and waste production more precisely and reduce the cost of reporting. This potential facilitates adherence to international ESG standards including the Global Reporting Initiative (GRI) and makes African businesses comply with the demands of investors to be transparent and responsible towards climatic conditions. With the AI system being integrated into operational processes, it will decrease the distance between the sustainability strategy and its implementation, making the environmental responsibility an ongoing, data-driven management process instead of a disclosure process that occurs periodically.

AIs also promotes the development and diffusion of new technologies and a continuous stream of innovation and knowledge that comes with it. African enterprises, for instance, have the ability to use AI tools for the data analysis and scenario simulation to cultivate new business models. AI fintech startups, for example, are designing advanced credit-scoring systems that broaden access to credit to many emerging markets, AI is also used by agritech startups to predict better and more efficiently to enhance food security and improved supply chain. These innovations are not just sectoral transformations, but also shifting the entire economy by fostering entrepreneurship, job creation, and technology transfer (AfDB, 2023).

Also, AI has improved regional cohesion and global competitiveness by allowing African businesses to participate in GVC digitally. Intelligent automation offers better compliance with international quality regulations, aids in cross-border e-commerce, and improves transparency throughout the logistics and trade integration processes. Analytics powered by AI and blockchain integrated systems for cross-border supply chain management and customs facilitation. This is the position of African firms in the global digital economy. African firms are no longer mere consumers of technology. They are producers of transformative technology on the global stage.

The need for AI on a national and continental level is a matter of strategy. They are equally essential for the economic and social returns that digital transformation brings. Countries on the continent like South Africa, Egypt, and Nigeria have implemented national strategies built around the ethical deployment of powerful AI which They have also positioned themselves to prosper from South Africa, Egypt, Nigeria, and Kenya have all initiated the construction of tailored national strategies focusing on the creation of powerful AI systems emphasized on data and innovation around ethical governance of AI. AI is a transformative technology. In Africa, automation of AI processes is a way to promote a nation's success in the global digital economy. AI helps firms break free from conventional economic growth. It does so by fostering innovation, structural environmentalism, and competitive resilience in the economy. How successful the continent's long-term success hinges on the balance between the foresighting policies of the continent on the driven transformation of AI technology and the inclusive developmental policies as a framework within which such developmental interaction takes place and finally, if the inclusive and equitable development being strived for will indeed result in developmental value being shared among businesses, the society and the generations to come.

5 Conclusions and Recommendations

5.1 Conclusions

The chapter concludes that AI workflow automation is a game-changing component for leapfrogging Africa towards sustained, tech-enabled growth on the continent. African enterprises are adopting AI technologies to streamline workflows, make data-driven decisions, and formulate scalable, inclusive, and resilient business models. However, sustained acceleration entails a purposeful alignment and balance between technological advancement, human capital investment, and the governance matrix. In the absence of ethical and capacity-imposing infrastructure, AI is likely to widen the prevailing inequalities and underutilize its transformative impact. This strategic alignment to ethical and equity frameworks for the application of AI technologies on the continent shall forge the future of African business ecosystems. This will require cross-sectoral collaboration between governments, the private sector, academia, and civil society.

5.2 Recommendations

As a way to maximize the sustainability effect of AI workflow automation, the governments and businesses in Africa should focus on implementing intelligent systems that explicitly aim at energy efficiency and carbon reduction and minimize waste. To combat AI growth contributing to the carbon footprint of the continent, renewable-powered data infrastructure and energy-saving cloud services should be invested in. On a firm level, the sustainability measures, i.e., the emissions intensity, resource efficiency, and waste reduction, need to be integrated into the automated work processes and decision-support systems, whereby the environmental performance goals are in tally with the cost and productivity ones. Policymakers are also advised to incentivize AI uses that facilitate green innovation in logistics, agriculture, energy management, and circular supply chains, even more, as well as increase the quality of ESG reporting and the standards of data governance. Such a strategy would make AI workflow automation a cornerstone of making Africa shift to inclusive, low-carbon, and robust economic development in the future when paired with workforce reskilling and ethical AI governance.

5.3 Limitations and Suggestions for Further Study

5.3.1 Limitations

While this chapter provides a thorough conceptual and empirical evaluation of AI workflow automation and its impact on sustainable business growth in Africa, there are some limitations that need to be addressed. The lack of primary data—such as structured interviews and original surveys—casts a shadow of a firm-level understanding of organizational contextual adoption patterns. The chapter is also confined temporally and geographically, as many data-gathering attempts synthesize data from various international regions and, therefore, lose country-specific insights. The lack of a universally accepted metric of sustainability also makes it exceedingly challenging to determine the precise environmental and social impact of AI, especially on MSMEs. Algorithmic discrimination, data privacy issues, and the impact of automation on employment are some of the ethical issues that had to be addressed at the conceptual level, as there was insufficient access to privacy-controlled datasets. Lastly, the predominance of Western literature on AI in use raises the possibility of neglecting the sociotechnical realities of Africa, stressing the importance of including Afrocentric approaches and local data in future scholarship to broaden the contextual and empirical scope and ensure adequate Africa-relevant research.

5.3.2 Suggestions for Further Study

While insights and limitations have been discussed and addressed, further valuable exploration is still needed with respect to AI's role in Africa's transformation to sustainable business. Firm-level, empirical studies are needed on the impact AI has on productivity, employment, and profitability, while regional studies on the differences in adoption due to the varying levels of infrastructure and differing policy settings would be useful. More research is needed on responsible AI in relation to local cultural norms, gender equity, and governance frameworks. More focus is needed on how affordable cloud-based AI technologies can be used by MSMEs to scale and become more competitive. Also, how AI assists in green transformation will help define its role in the low-carbon development and sustainable development goals. It is equally important to assess the impact of AI on the labor market, skill and wage structures, and other longitudinal studies on the adoption of AI and policy frameworks by 2030.

References

African Development Bank (AfDB). (2023). *Digital transformation strategy for Africa (2020–2030)*. AfDB.

African Union (AU). (2022). *AI for Africa: Policy framework for responsible artificial intelligence*. AU Commission.

African Union (AU). (2022). *Data policy framework for Africa*. African Union Commission.

Brynjolfsson, E., & McAfee, A. (2021). *The business of artificial intelligence: What it can and cannot do for your organization*. Harvard Business Review Press.

Carroll, A. B., & Brown, J. A. (2018). Corporate social responsibility: A review of current concepts, research, and issues. *Business & Society, 57*(1), 7–23.

Cascio, W. F., & Boudreau, J. W. (2019). *Investing in people: Financial impact of human resource initiatives* (3rd ed.). Pearson.

Crawford, K. (2021). *Atlas of AI: Power, politics, and the planetary costs of artificial intelligence*. Yale University Press.

Creswell, J. W., & Poth, C. N. (2018). *Qualitative inquiry and research design: Choosing among five approaches* (4th ed.). SAGE Publications.

Davenport, T. H., & Ronanki, R. (2022). *Artificial intelligence for the real world*. Harvard Business Review Press.

Dyllick, T., & Muff, K. (2016). Clarifying the meaning of sustainable business: Introducing a typology from business-as-usual to true business sustainability. *Organization & Environment, 29*(2), 156–174.

Elkington, J. (1997). *Cannibals with forks: The triple bottom line of 21st century business*. Capstone Publishing.

Eubanks, V. (2018). *Automating inequality: How high-tech tools profile, police, and punish the poor*. St. Martin's Press.

Freeman, R. E., Harrison, J. S., & Zyglidopoulos, S. (2018). *Stakeholder theory: Concepts and strategies*. Cambridge University Press.
Hart, S. L., & Dowell, G. (2011). Invited editorial: A natural-resource-based view of the firm—Fifteen years after. *Journal of Management*, *37*(5), 1464–1479.
International Labour Organization (ILO). (2022). *The future of work in Africa: Harnessing AI for inclusive growth*. ILO.
KPMG. (2022). KPMG Global *AI in finance Report*. KPMG Nigeria.
McKinsey & Company. (2023). *The economic potential of generative AI: The next productivity frontier*. McKinsey Global Institute.
Organisation for Economic Co-operation and Development (OECD). (2023). *OECD framework for the classification of AI systems*. OECD Publishing.
Organisation for Economic Co-operation and Development (OECD). (2023). *Green growth and sustainable development 2023*. OECD Publishing.
Pasmore, W. (2015). *Research in organizational change and development: Socio-technical systems theory in practice*. Emerald Group Publishing.
Porter, M. E., & Kramer, M. R. (2019). Creating shared value: Redefining capitalism and the role of the corporation in society. *Harvard Business Review*, *97*(1), 65–77.
PwC. (2022). *AI and automation in Africa: Bridging the productivity gap*. PwC Africa Insights Report.
Rogers, E. M. (2003). *Diffusion of innovations* (5th ed.). Free Press.
Teece, D. J. (2018). *Dynamic capabilities and (digital) platform lifecycles. Industrial and Corporate Change*, *27*(4), 1–21.
Trist, E., & Bamforth, K. (1951). Some social and psychological consequences of the longwall method of coal-getting. *Human Relations*, *4*(1), 3–38.
UNESCO. (2023). *AI skills readiness index: Building human capacity in Africa*. UNESCO.
United Nations Conference on Trade and Development (UNCTAD). (2023). *Digital economy report 2023: Value creation and capture*. UNCTAD.
World Bank. (2024). *Africa's digital economy: Catalyzing inclusive growth through AI and automation*. World Bank Group.
World Economic Forum (WEF). (2023). *The future of jobs report 2023*. World Economic Forum.
Yin, R. K. (2018). *Case Study Research Design and Methods* (6th Ed.). Thousand Oaks, CA: Sage Publishing.

Chapter 3
AI-Driven Entrepreneurship: Catalyzing Green and Inclusive Growth in Africa

Dennis I. Edogun

Abstract This chapter examines the link between technological innovation and sustainable development in Africa. While artificial intelligence (AI) often receives considerable praise for its potential to disrupt industries, its role as a purposeful driver of green and inclusive structural change hasn't been fully explored and is often hindered by systemic challenges. The chapter highlights AI-driven entrepreneurship, the thoughtful integration of AI into business models that tackle social and environmental issues, as a crucial tool for pushing forward the United Nations Sustainable Development Goals (SDGs) and creating a fresh approach to inclusive capitalism across the continent. Using a qualitative analytical framework that draws on insights from case studies, sector reports, and academic literature spanning technological innovation, sustainability science, and development economics, the chapter tackles two main questions: How are AI-enabled businesses contributing to green innovation and inclusive business practices across key African sectors? And what are the main structural, ethical, and ecosystem challenges they encounter? The findings show that AI is playing an active role in driving structural transformation, from precision agriculture that enhances climate resilience and boosts smallholder incomes to off-grid renewable energy solutions and inclusive fintech platforms that broaden access for underserved communities. While these exciting applications hold great promise, they face a "dual gap" in both foundational data and skills. This issue is further complicated by shortcomings in digital infrastructure, mismatched funding, and the ongoing threat of "algorithmic apartheid," which can perpetuate existing inequalities. As a result, the chapter makes a compelling case that, to unlock AI's potential for sustainable development truly, we need to shift from the idea of "AI for Africa" to "AI by and with Africa." Its unique contribution lies in weaving together these complex challenges into a cohesive, ecosystem-based strategy, suggesting practical ways to foster context-specific innovation through strategic partnerships between the public and private sectors, flexible policy frameworks, and

---

D. I. Edogun
Conestoga College, Canada
e-mail: denogun@yahoo.com

K. Modugu, J. Otalor (eds.), *Sustainability and Digital Transformation in Africa*, Frontiers in African Business Research,
https://doi.org/10.1007/978-981-95-9500-6_3

financing linked to impact. Ultimately, the chapter offers evidence-based insights for key players: for policymakers, it emphasizes the importance of AI governance that focuses on data sovereignty and ethical standards; for entrepreneurs and investors, it points out the business potential of green, inclusive models; and for sustainability advocates, it positions AI-driven entrepreneurship as a vital means to achieve fair and resilient development throughout Africa.

Keywords Artificial intelligence · sustainable development · inclusive entrepreneurship · Africa, digital innovation · green growth · policy · ecosystem approach

1 Introduction

A new chapter is unfolding in Africa, and it's not just about the usual commodities anymore; it's about the remarkable potential of artificial intelligence (AI). At the lively intersection of AI, sustainable development, and inclusive entrepreneurship, we're witnessing an opportunity for Africa to take its most significant economic leap yet. This isn't some distant fantasy; AI is rapidly transitioning from a theoretical concept to a crucial element of Africa's economic landscape, poised to redefine its role on the global stage. The numbers are impressive: projections indicate that AI could double Africa's GDP growth rate by 2035, fundamentally changing its economic path (Ifeanyi et al., 2025; Deshpande, 2025). By 2030, this technological wave is expected to add an astonishing USD 1.5 trillion to the African economy; a potential that hinges on the continent effectively engaging with the global AI revolution and claiming its rightful place in the digital future (Deshpande, 2025).

But beyond those eye-catching economic figures, the true promise of AI lies in its capacity to foster human progress. This analysis underscores AI as a pivotal force in achieving the United Nations Sustainable Development Goals (SDGs), transforming computational power into tangible, life-changing results. We're already witnessing how AI applications align with Africa's most pressing challenges: from precision agriculture aimed at Zero Hunger (SDG 2) and diagnostic tools that enhance Good Health and Well-being (SDG 3), to smart grids that improve Affordable and Clean Energy (SDG 7) and predictive models that bolster Climate Action (SDG 13) (Ifeanyi et al., 2025; APET Report, 2024; Deshpande, 2025). This is where technology meets purpose, delivering not just efficiency, but also equity and sustainability.

This chapter dives deep into an exciting new trend: AI-driven entrepreneurship. It's not just about using the latest software; it's about a strategic, goal-oriented approach where AI is central to business models designed to address environmental and social challenges (Ifeanyi et al., 2025). This movement is all about innovators and founders who are building businesses that link profit-making with creating a positive impact. In the following analysis, the author will thoroughly investigate this emerging ecosystem, examining innovative applications, navigating complex challenges, and outlining the essential steps to ensure that Africa's AI revolution is both inclusive and transformative.

2 Literature Review

This chapter lays down a theoretical foundation by critically engaging with three interconnected areas of literature that inform the study of AI-driven entrepreneurship in Africa: (1) a critique of techno-utopianism and the political economy of technology; (2) theories of sustainable development and inclusive innovation; and (3) the emerging scholarship on decolonial and situated AI. By synthesizing these perspectives, we gain the conceptual tools needed to move beyond deterministic hype and analyze AI as an embedded socio-technical phenomenon with varied developmental outcomes.

2.1 Beyond Technosolutionism: AI in the Political Economy of Development

A crucial starting point is a critical examination of "technosolutionism," the idea that technology alone can resolve complex socio-political issues (Morozov, 2014). In the context of Africa, this often appears as technological determinism, where AI is portrayed as a decontextualized "savior" of systemic problems in governance, infrastructure, and service delivery (Okolo, 2024). This narrative often gets a boost from a political economy in which global tech companies and investors, eager to tap new markets, have a strong interest in promoting AI as a must-have solution.

This can sometimes overshadow local needs and the importance of evidence-based assessments. This critique pushes us to look beyond just the technical aspects of algorithms and to consider who is creating AI, for whom it's being created, and what the consequences are: questions that are crucial when viewed through a political economy lens (Foster & Azmeh, 2020).

2.2 AI as a Catalyst: Linking to Sustainable Development and Inclusive Entrepreneurship

To frame AI as a potential driver for positive change, this chapter leans on theories of sustainable development and inclusive innovation. Sustainable development, especially as outlined in the UN SDGs, offers a guiding framework for assessing the impact of technology based on environmental sustainability, social inclusion, and economic fairness (UN, 2015). At the same time, the discussion around inclusive innovation examines how entrepreneurship can be tailored to meet the needs of marginalized groups (Foster & Heeks, 2013). By bringing these ideas together, we define AI-driven entrepreneurship as the thoughtful use of AI technologies within business models intentionally designed to address socio-environmental market failures while generating measurable impacts alongside financial gains. This shifts the focus from whether AI is being used to how its application aligns with the principles of green growth and equitable access.

2.3 *Situating AI: Decolonial Perspectives and Contextual Relevance*

The last theoretical pillar delves into decolonial AI and situated knowledges scholarship, which argues that technology isn't one-size-fits-all but rather reflects the values and contexts of those who create it (Birhane, 2020; Mohamed et al., 2020). This body of work highlights the dangers of "algorithmic colonialism," in which systems developed elsewhere extract data and impose foreign logic, potentially perpetuating historical inequalities (Couldry & Mejias, 2019).

A significant issue we face is data injustice stemming from datasets that don't represent everyone, particularly the marginalization of African languages and ways of knowing in AI development (Aigbiniode, 2025). This perspective highlights the need for relevance in context, advocating technical approaches such as federated learning and small language models that can adapt to local needs (Aigbiniode, 2025), as well as design processes that involve local communities in meaningful ways.

2.4 *Synthesizing the Framework: Towards an Analytical Lens*

By weaving together three key areas, the critique of political economy, the ethical goals of sustainable and inclusive development, and the decolonial push for context-aware design, this chapter builds a comprehensive analytical framework. This framework doesn't dictate how AI will affect us but instead offers the conceptual tools needed to critically assess: how well AI initiatives align with the values of the Sustainable Development Goals (SDGs); the structural ecosystems that either support or hinder these initiatives; and the ethical and knowledge practices that ensure innovations are fair and grounded in local realities. This synthesis directly shapes the methodological approach for a systematic thematic analysis, steering the exploration of sector-specific applications, structural challenges, and ethical concerns that will follow.

3 Methodology

This chapter adopts a structured, qualitative methodology to conduct a thorough literature review and a critical thematic synthesis. The goal is to develop a robust conceptual framework for understanding how AI-driven entrepreneurship can catalyze sustainable and inclusive development in Africa. The methodology is carefully crafted to guarantee transparency, reliability, and academic rigor through a replicable three-phase process.

3.1 Systematic Literature Identification and Selection

To build a solid foundation of evidence, we carried out a thorough search across four key scholarly databases: Scopus, Web of Science, IEEE Xplore, and Google Scholar, along with gray literature from critical international organizations like the World Bank, AfDB, UN agencies, and GSMA. We crafted search strings that combined terms related to ("artificial intelligence" OR "machine learning"), ("entrepreneurship" OR "startup" OR "innovation"), ("Africa" OR "Sub-Saharan Africa"), and ("sustainable development" OR "inclusive growth" OR "SDGs").

The initial collection was filtered using specific inclusion and exclusion criteria: sources had to (a) focus significantly on AI applications within an African context, (b) address entrepreneurial, developmental, or policy aspects, and (c) be published in English between 2013 and 2025. We chose industry case studies (such as Zencey and Jumo) for in-text citations based on their representativeness, documented impact, and frequent mention across various reports as examples of applied AI innovation.

3.2 Analytical Framework and Thematic Synthesis

The literature we selected was examined through a structured qualitative thematic analysis, guided by a preliminary conceptual framework that emerged from the intersections of technological innovation, sustainability science, and development economics. This process included:

1. Getting Acquainted and Open Coding: We kicked things off with a thorough read-through of the literature, coding it to highlight key concepts, which we then organized in a digital coding matrix.
2. Axial Coding and Theme Development: From those initial codes, we carefully grouped and refined them into three main and interconnected analytical themes that guide the chapter's exploration:

- Sectoral Applications and Value Creation: This theme examines how and where AI creates genuine economic, social, and environmental value, such as in agritech, fintech, and healthtech.
- Structural and Ecosystem Barriers: Here, we identify the material, technical, and institutional challenges, like data gaps, skill shortages, and funding models, that hinder the growth of AI ventures.
- Ethical Risks and Responsible Innovation: We take a critical look at issues such as bias, privacy, equity, and governance to ensure our framework is grounded in the principles of responsible AI.

3. Constant Comparative Analysis: We continuously compared insights from academic literature, case studies, and policy reports within and across these themes to spot patterns, contradictions, and gaps in evidence.

3.3 Synthesis and Conceptual Framework Development

In the final phase, we synthesized our thematic analysis into a central conceptual framework for the chapter. This wasn't just a straightforward summary; it was an analytical effort to clarify the dynamic relationships among the sectoral opportunities, structural challenges, and ethical responsibilities we identified. We developed and validated this framework through an iterative process, mapping out these relationships to ensure that our model clearly outlines the pathways and prerequisites for AI-driven entrepreneurship to foster green and inclusive structural transformation in Africa.

This systematic, multi-phase approach guarantees that our findings are based on a transparent and critically evaluated evidence base. We go beyond just summarizing narratives to lay down a solid analytical foundation for our conclusions and recommendations. The methodology we used clearly outlines how we handled the literature and constructed our themes, addressing concerns about replicability and enhancing the scholarly impact of our analysis.

4 Findings and Discussion from a Thematic Analysis of AI-Driven Entrepreneurship in Africa

In this chapter, we dive into the key findings from our systematic thematic analysis, which is organized around three interconnected themes that emerged from the data. We go beyond just describing the results; we also show how the methodology outlined in Section 3 led us to these insights. Each subsection kicks off with a summary of the theme that emerged from our coding process, blends evidence from existing literature and case studies, and wraps up with a discussion that ties these findings back to the theoretical framework we set up in Section 2. This approach provides a transparent and credible link between our methods and our analytical claims.

4.1 *Sectoral Applications and Value Creation for Sustainable Development*

Through open and axial coding of academic literature, industry reports, and case studies, we uncovered a central theme of applied value creation. This theme highlights how AI is being practically woven into entrepreneurial efforts to create tangible economic, social, and environmental impacts that align with specific Sustainable Development Goals (SDGs). Our analysis pinpointed three key sectoral clusters where this value creation is particularly evident and well-documented.

4.1.1 Precision Agriculture and Climate Resilience (Aligning with SDG 2 and 13)

Analytical Finding: A review of various agri-tech case studies (such as CropScan and Zenvus) and sector reports reveals a clear trend: AI is being used specifically to help smallholder farmers adapt to climate change and build resilience (Olatunbosun, Adeyemi, & Nwankwo, 2026). The real value lies in creating hyper-local, plot-specific insights by combining IoT, satellite, and weather data through machine learning (Muhammed et al., 2024). This innovative approach enables targeted actions that reduce input costs, minimize crop losses, and encourage sustainable land practices. The result? Improved economic viability for farms and better environmental care as we face climate challenges (World Bank, 2022).

Evidence Synthesis: The analysis brought together startup whitepapers, investor insights, and academic evaluations to outline how AI is shaping precision agriculture. It highlighted a common approach: using affordable IoT sensors and satellite imagery to monitor field and crop conditions (Olatunbosun, Adeyemi, & Nwankwo, 2026). Machine learning processes this information to send pest alerts, suggest optimal planting times, and create tailored irrigation plans, all accessible to farmers via user-friendly mobile platforms (SMS and apps). This effectively connects data science with real-world farm management (Muhammed et al., 2024).

The gathered evidence shows a clear connection to tangible results. Documented pilot projects indicate yield increases of 20%–30% for staple crops thanks to optimized inputs (World Bank, 2022) and about a 15% reduction in water usage, which is a significant advantage in drought-prone areas (Bekele-Thomas, 2024). These findings confirm that AI is moving from theory to real-world application, providing measurable improvements in productivity and sustainable resource management, which are crucial for food security and climate resilience (Tsan et al., 2019).

Discussion: This finding really taps into the heart of the chapter's main ideas about inclusive innovation and sustainable development (Foster & Heeks, 2013; UN, 2015). By customizing IoT and machine learning to meet the needs of smallholder farmers, these initiatives showcase what we call "frugal innovation." They're crafting affordable and sturdy solutions that support various Sustainable Development

Goals (SDGs), such as productive agriculture (SDG 2) and climate action (SDG 13) (World Bank, 2022).

However, the analysis highlighted a vital point: scalability often hinges on additional, non-technological factors, such as dependable mobile connectivity and farmers' digital literacy (Olatunbosun, Adeyemi, & Nwankwo, 2026). This highlights the deep connection between sectoral opportunities (Theme 1) and structural challenges (Theme 2). If we focus solely on AI applications without also investing in rural broadband and skill development, we risk creating isolated successes that can't achieve transformative change on a continent-wide scale. This underscores the need for integrated strategies that consider the entire ecosystem.

4.1.2 Inclusive Fintech and Economic Agency (Aligning with SDG 1, 8, and 10)

Analytical Finding: A thematic analysis of the fintech literature and the business models of leading platforms such as Jumo and Branch identified a recurring theme: the smart use of artificial intelligence for flexible, alternative credit assessments (Abiero, 2024; Suri & Jack, 2016). This method employs machine learning algorithms to analyze nontraditional data sources, such as mobile money transaction histories, airtime purchase behaviors, smartphone usage patterns, and even social network metrics. This helps create a real-time behavioral profile of creditworthiness for individuals who are often excluded from formal banking systems (Björkegren & Grissen, 2020; Cook & McKay, 2015).

This fresh approach offers significant value through two key avenues. First off, it helps pull a substantial amount of informal economic activity into the financial system by transforming everyday digital footprints into a clear financial identity. This integration is crucial for bringing marginalized workers into the formal digital economy (Barboni et al., 2018). Secondly, and just as vital, it serves as an essential safety net for micro-entrepreneurs and gig workers by providing them with quick, small loans for emergencies, inventory needs, or to help them get through tough times. This not only enhances their economic resilience but also opens up new growth opportunities (Munyegera & Matsumoto, 2016).

Evidence Synthesis: When it comes to evidence synthesis, the analysis combined academic studies, corporate impact reports, and user testimonials to confirm the effectiveness of AI in promoting financial inclusion. It showed that AI-driven risk models that analyze alternative data from mobile money, airtime, and smartphone usage can create a reliable "digital identity" for those typically excluded from traditional banking. This allows for accurate credit assessments without the need for conventional collateral (Björkegren & Grissen, 2020; Suri & Jack, 2016).

The synthesized evidence highlights a multi-faceted value proposition. The main benefit is increased access to instant capital, which helps smooth consumption and enables business investments (Barboni et al., 2018). This process builds a digital financial history ("credit-building"), allowing users to eventually qualify for better terms (Munyegera & Matsumoto, 2016). Additionally, the rich behavioral data collected enables customized micro-insurance options (like health or crop insurance), which are crucial for building resilience and creating a sustainable model for financial inclusion (Dercon & Kirchberger, 2021).

Discussion: The use of AI in inclusive fintech aligns with the inclusive capitalism aspect of the theoretical framework, showcasing how market-based innovations can effectively tackle systemic financial exclusion and empower marginalized economic participants (Foster & Heeks, 2013). By tapping into alternative data sources to provide credit to individuals without bank access, these platforms enable economic advancement, fostering micro-entrepreneurship and helping establish digital financial identities (Suri & Jack, 2016; Barboni et al., 2018).

Yet, when viewed through a political economy lens, a significant tension within this model emerges (Foster & Azmeh, 2020; Zuboff, 2019). While they democratize access, these platforms risk creating a massive concentration of sensitive financial and behavioral data within the proprietary algorithms of a handful of private, often foreign-owned, tech companies. This situation raises serious questions about data sovereignty, who really owns, controls, and benefits from this collective digital footprint, and the potential for long-term market power consolidation that could hinder competition (Couldry & Mejias, 2019). This issue directly connects to the ethical concerns in Theme 3, where the risks of algorithmic bias and privacy erosion arise from this concentrated control over essential economic data and decision-making systems (Erastus, 2021; Abiero, 2024).

4.1.3 Decentralized Access in Healthcare and Education (Aligning with SDG 3 and 4)

Analytical Finding: A thorough analysis of health-tech and ed-tech initiatives across Africa uncovered a clear and recurring trend: AI is being used strategically as a force multiplier to address critical human resource shortages while adding a new layer of personalization to service delivery (Ifeanyi et al., 2025; Abiero, 2024). In healthcare, AI-driven tools enhance clinicians' reach by triaging cases and equipping community health workers with improved, protocol-based guidance, rather than replacing them (Winters et al., 2018). In education, adaptive learning platforms adjust content and pacing to meet individual student needs, creating a personalized learning experience in overcrowded classrooms where one-on-one teacher interaction is often unfeasible (Asiegbu & Okolo, 2024).

The real value of these innovations lies in their ability to democratize access. They use algorithms to bridge gaps in geography and resources, making quality diagnostic services and tailored educational instruction available to groups that have often been left out due to systemic inequalities (Ogundipe et al., 2023).

Evidence Synthesis: A systematic review of telemedicine platforms, like Zencey, and adaptive learning tools looked into their effectiveness in low-resource environments. The findings pointed to a key insight: these technologies are meant to enhance, not replace, human professionals. AI symptom checkers help triage patients and support overwhelmed staff, while mobile tools offer protocol-driven decision support for community health workers (Winters et al., 2018). In the education sector, adaptive software personalizes learning for large classrooms where individual teacher attention is often lacking (Majgaard & Misfeldt, 2021).

Quantitative data support improvements in access and quality. Telemedicine expands geographic reach, allowing remote patients to receive specialist care (Molli, 2021), while AI-driven reminders enhance service adherence. In classrooms, adaptive tools increase engagement and foundational skills, particularly for students who are struggling or excelling (Burns, 2021). The evidence suggests that when thoughtfully integrated as a supportive layer within human-centered systems, AI can truly democratize access and improve the effectiveness of essential services, tackling fundamental inequities.

Discussion: The fields of health and education highlight the ongoing struggle between the excitement of technological solutions and the realities on the ground (Timcke, 2024). Recent case studies and reviews indicate that the most successful AI models are those that function as human-in the-loop systems. These models enhance, rather than replace, the essential judgment, empathy, and relational roles that nurses and teachers play (El Arab et al., 2025; International Task Force on Teachers for Education 2030, 2025).

This finding strongly supports the decolonial AI focus on situated design (Mohamed et al., 2020; Birhane, 2020). Successful innovation tailors' technology to fit into existing human systems, local workflows, and community knowledge. For instance, a symptom-checker needs to seamlessly integrate into a health worker's daily routine, while a learning platform should resonate with a teacher's teaching style. This idea of situatedness ensures that technology is adaptable to its context, protecting local expertise and prioritizing human dignity and agency (Abiero, 2024; Ifeanyi et al., 2025).

4.2 Structural and Ecosystem Barriers to Scaling Innovation

By analyzing policy documents, interviewing founders, and reviewing infrastructure reports, a recurring, interconnected theme emerged: structural constraints. This theme underscores the material, technical, and institutional challenges that consistently hinder the growth and sustainability of AI initiatives, regardless of their potential across sectors.

4.2.1 The Foundational Dual Gap: Data Scarcity and Skill Shortages

Analytical Finding: The analysis reveals a crippling and self-reinforcing dual constraint at the core of Africa's AI landscape: a significant "data desert" characterized by limited, isolated, and unrepresentative datasets exists alongside a critical "talent desert" that faces a severe shortage of skilled data scientists, machine learning engineers, and AI ethicists (Research ICT Africa, 2023). This scarcity isn't just a random occurrence; it's woven into the fabric of the system. The lack of large-scale, open, and well-organized data poses a significant obstacle for training effective AI models and for adequately educating new talent. Additionally, the absence of a robust pool of local experts complicates the establishment of the advanced data infrastructure and governance frameworks needed to develop high-quality data assets (Gitau, 2025).

This creates a challenging cycle: without quality data, we struggle to train or utilize talent effectively, and without skilled individuals, we can't generate, curate, or fully leverage valuable data. This cycle really hampers innovation, stifles the growth of AI-driven businesses, and prevents the entire ecosystem from developing homegrown, contextually relevant solutions, leaving us trapped in a cycle of technological dependency (Foster & Azmeh, 2020).

Evidence Synthesis: The analysis of policy briefs from sources like Research ICT Africa highlights the continent's foundational "data desert": a significant lack of large-scale, curated, open datasets due to fragmented collection efforts, restrictive governance, and data siloing in both public and private sectors. Representation issues further complicate this scarcity, as existing datasets often overlook Africa's diverse linguistic and socioeconomic realities, resulting in a shortage of high-quality, locally relevant data necessary for robust AI (Research ICT Africa, 2023; Ubalijoro et al., 2023).

At the same time, examining human capital reports and university programs reveals a parallel "talent desert." There's a critical shortage of advanced data scientists and machine learning engineers, worsened by limited specialized AI curricula and the phenomenon of "brain drain," where top graduates are recruited globally, draining the local talent pool that's essential for innovation and developing data solutions (Gitau, 2025; Ndung'u & Signé, 2020).

Discussion: This observation sheds light on the political economy critique, illustrating that the "data desert" and "talent desert" are products of systemic power imbalances, not mere coincidence (Foster & Azmeh, 2020; Zuboff, 2019). The data gap arises from underinvestment in public data infrastructure and from a global digital economy in which multinationals extract raw data from African users, leaving behind few locally owned datasets (Couldry & Mejias, 2019). The talent gap is worsened by brain drain, as top graduates are recruited from around the world, depleting the local talent pool (Ndung'u & Signé, 2020). This twofold extraction process seriously undermines local innovation.

To tackle this issue, we need comprehensive, nation-led strategies that go beyond just holding isolated workshops. One practical approach could be to create national data trusts, which are ethical, sovereign repositories that prioritize public-interest innovation. This would help shift control back to the public domain from private entities (Research ICT Africa, 2023). At the same time, it's crucial to establish regional AI centers of excellence that offer strong retention incentives, such as competitive funding and clear industry pathways, to attract, nurture, and retain top talent in the area (Gitau, 2025). Only by implementing these integrated, structural strategies can we break the cycle of extraction and foster a self-sustaining ecosystem.

4.2.2 Capital Constraints and Misaligned Funding Models

Analytical Finding: Taking a closer look at financing literature uncovers a notable gap between the capital available and what entrepreneurs actually need, especially in the realms of deep-tech and sustainable businesses. Research indicates that while funding for early-stage companies is on the rise, it's still largely funneled into just a handful of sectors and regions. This leaves a significant void for innovative, capital-heavy models (Gornall & Strebulaev, 2020; World Bank, 2022). This disconnect leads to a challenging "valley of death" for startups that need substantial, patient, and risk-tolerant investments to transition from pilot projects to full-scale operations; a financing hurdle that's been well-documented in emerging innovation ecosystems (Acemoglu et al., 2022; Lerner, 2020).

This gap is especially pronounced for ventures that need to fine-tune hardware-software integration, navigate complex regulations (like those in healthtech), or develop markets in underserved regions where returns can take years to materialize (GSMA ClimateTech, 2023; Agwaibor, 2025). This misalignment diverts innovation away from urgent structural challenges, underscoring the need for catalytic public funding, blended finance, and venture capital with longer timelines to de-risk these high-impact, capital-intensive pathways.

Evidence Synthesis: When we look at the investment landscape and gather insights from founder surveys, it becomes clear that there's a strong bias among investors towards capital-light, software-only "applied AI" in established sectors like fintech. These areas promise quicker exits and align with the typical risk-return profiles that investors are comfortable with (Agwaibor, 2025; Gornall & Strebulaev, 2020; World Bank, 2022). As a result, much funding is directed toward digitizing existing processes rather than nurturing groundbreaking innovations.

This trend sharply contrasts with what's happening in climate tech, biotech, and deep-tech fields. For instance, case studies in precision agriculture and renewable energy show that these sectors often struggle to attract the patient, growth-stage capital they need to move from pilot projects to full-scale commercial operations. This leads to what's been termed "pilot purgatory" (Gornall & Strebulaev, 2020; World Bank, 2022; GSMA ClimateTech, 2023). Consequently, these high-impact industries, which are essential to tackling structural challenges and promoting green industrialization, face a daunting "valley of death." This is primarily due to the mismatch between their lengthy development timelines and the shorter investment horizons that many investors prefer.

Discussion: This scenario underscores a significant conflict between the long-term objectives of sustainable development and the short-term expectations of private capital regarding risk and return (Foster & Azmeh, 2020; Mazzucato, 2021). While the Sustainable Development Goals (SDGs) call for patient investment in systemic solutions like climate tech, the market tends to favor quick returns in established, lower-capital sectors like fintech. This creates a "valley of death" for high-impact, capital-intensive projects (Gornall & Strebulaev, 2020; World Bank, 2022; GSMA ClimateTech, 2023).

Now, the newly announced $60 billion Africa AI Fund is being touted as a potential solution to bridge this financing gap (Munga & Quansah, 2025). However, its effectiveness will hinge on its mandate and governance (Agwaibor, 2025). To truly make a difference, it needs to focus on providing catalytic patient capital for high-impact, high-risk ventures in areas like climate-smart agriculture and inclusive health tech, measuring success by sustainable development outcomes. If it doesn't adopt this impact-driven approach, there's a real risk that the fund could end up reinforcing the very investment biases it aims to correct.

4.2.3 Infrastructure Deficits and Regulatory Uncertainty

Analytical Finding: Our thematic analysis has uncovered a vital set of material and institutional supports that are often lacking, unreliable, or underdeveloped in many African economies, posing a significant hurdle to AI-driven entrepreneurship. This trio of essential supports includes, first and foremost, the need for stable, widespread electricity; without it, data centers, edge devices, and digital workspaces can't operate. Next, we need affordable, high-speed internet access, which is crucial for cloud computing, real-time data sharing, and connecting to global platforms and knowledge resources. Lastly, having precise, predictable, and supportive regulatory frameworks that specifically address AI deployment, data governance, and digital commerce is equally essential.

These frameworks provide the legal certainty needed to attract substantial investments and scale operations. The ongoing absence or weakness of these supports in policy documents and startup narratives indicates that they aren't just minor issues; they are fundamental to the ecosystem's health and the types of innovations that can thrive sustainably.

Evidence Synthesis: The analysis brought together technical reports on grid and broadband reliability (World Bank, 2022), legal analyses (State of Internet Freedom, 2025), and startup case studies, highlighting how frequent outages and high connectivity costs pose significant operational challenges (Mawadza & Burton, 2023). This multi-source evidence confirms that infrastructure directly influences the viability of AI business models. The high costs and latency associated with cloud computing, coupled with intermittent connectivity, make cloud-reliant solutions unstable, particularly for real-time or rural applications.

As a result, the analysis reveals a straightforward market adaptation: entrepreneurs are increasingly incentivized to lean towards edge-based computing solutions (State of Internet Freedom, 2025; Oughton et al., 2021; Cassava Technologies, 2023). By processing data locally, these models reduce the need for constant bandwidth and lower ongoing cloud expenses, making them more robust in today's infrastructure landscape. This shift highlights that infrastructure actively shapes innovation rather than just serving as a passive backdrop.

Discussion: These gaps represent classic institutional voids; systemic failures in foundational structures and rules (Okolo, 2024). The lack of electricity and broadband is a clear sign of historical underinvestment, often overlooked amid the hype surrounding technological solutions, and it directly influences innovation by promoting resilient, edge-based solutions rather than those reliant on the cloud (Okolo, 2024; Foster & Azmeh, 2020).

On the flip side, regulatory uncertainty offers a unique opportunity for proactive governance (Ogundipe et al., 2023; Bernard, 2024). Unlike established markets that must retrofit existing rules, African nations have the opportunity to create context-appropriate, innovation-friendly frameworks from scratch, embedding ethical principles such as fairness and accountability from the start (Abiero, 2024). This approach can encourage responsible innovation that aligns with local values and supports sustainable development, potentially steering clear of the regulatory hurdles faced in other regions.

4.3 Ethical Risks and the Need for Responsible Innovation

As we explored the coding process, a recurring theme of socio-technical risk emerged, threading through discussions across sectors and structures. This theme underscores the unintended consequences and power imbalances that can arise when AI is implemented in complex and unequal societies.

4.3.1 Algorithmic Bias and the Risk of "Algorithmic Apartheid"

Analytical Finding: One of the most significant ethical issues that came up in our analysis is the undeniable evidence showing that AI systems can, and often do, embed, perpetuate, and even automate deeply ingrained social biases (Sjoding et al., 2020; Wong et al., 2021; Obermeyer et al., 2019). This isn't just a future concern; it's a current reality, backed by forensic analyses of AI systems already operating in various sectors. For instance, studies on healthcare algorithms have revealed that training data biased towards lighter-skinned individuals can result in racially skewed diagnostic errors in tools like pulse oximeters (Sjoding et al., 2020; Letswado, 2025).

Similarly, audits of financial technology and recruitment platforms have uncovered patterns of discrimination that consistently disadvantage certain demographic groups (Letswado, 2025; Luikham, 2025). These examples are clear proof that, without deliberate and comprehensive intervention, AI can easily become a powerful tool for digitally reproducing and amplifying existing societal inequalities, turning implicit biases into explicit, systemic outcomes.

Evidence Synthesis: The analysis thoroughly reviewed studies on racially biased medical diagnostics and discriminatory algorithms in finance. In healthcare, research on pulse oximeters has definitively shown that their algorithms, primarily calibrated on lighter-skinned populations, deliver less accurate readings for patients with darker skin, risking dangerous undertreatment and highlighting how non-representative data embeds historical disparities into digital tools (Sjoding et al., 2020; Wong et al., 2021; Obermeyer et al., 2019).

At the same time, research and investigative efforts have revealed that financial and insurance algorithms often unfairly target marginalized communities for extra scrutiny, reflecting long-standing patterns of profiling. These instances underscore how seemingly neutral, "objective" algorithmic decision-making can lead to biased outcomes, such as unwarranted investigations and limited access to essential services (Benjamin, 2019; Eubanks, 2018).

Discussion: This finding strongly supports the warnings from decolonial AI scholars that technology is never truly neutral; the contexts of its creators influence it (Mohamed et al., 2020; Birhane, 2020; Nothias, 2025). The documented biases in healthcare and finance provide concrete evidence that AI systems can absorb and perpetuate societal hierarchies, serving as mechanisms for the digital reinforcement of structural inequality (Benjamin, 2019; Couldry & Mejias, 2019). This could lead to a new, coded frontier for discrimination (Erastus, 2024).

As a result, addressing these issues can't just be an afterthought; it's essential for the health and legitimacy of the ecosystem (Foster & Azmeh, 2020; UNCTAD, 2021). This calls for fundamental changes: fostering diverse, interdisciplinary design teams; establishing independent algorithmic audits; and embracing fairness-by-design principles that ensure equity from data collection to model objectives (Abiero, 2024; Raji et al., 2020). Without this comprehensive approach, the ecosystem risks being built on a foundation of embedded bias, which could jeopardize its social license and potential for inclusivity.

4.3.2 Data Privacy, Surveillance, and Asymmetric Governance

Analytical Finding: A clear and concerning pattern emerged from the analysis, revealing a significant governance imbalance in the continent's digital regulatory landscape (State of Internet Freedom, 2025; Bernard, 2024). On the one hand, there's notable progress in developing consumer data protection frameworks, with over 36 African nations enacting laws grounded in the principles of informed consent and data subject rights, as highlighted in the African Union's Malabo Convention (Olatunbosun, Adeyemi, & Nwankwo, 2026). However, this progress starkly contrasts with a significant lack of adequate constraints and transparent oversight regarding the surveillance powers of state actors and large corporations (Abiero, 2024).

The swift rise of technologies like facial recognition in public areas, tools for monitoring social media, and the extensive data gathering by both governments and private companies often happens in a legal gray area or under laws that just don't fit the digital world (Privacy International, 2021; World Bank, 2021). This mismatch in regulations creates a troubling imbalance: while rules for protecting commercial data are improving, individuals are facing increasing threats to their personal freedom and civic space due to rampant, unchecked surveillance by both the state and corporations. This kind of environment stifles free speech, peaceful protests, and political dissent, ultimately undermining the core foundations of a democratic digital society (Feldstein, 2019; Land, 2020).

Evidence Synthesis: The analysis highlighted two simultaneous trends in Africa's digital governance: the encouraging spread of data protection laws, spurred by the Malabo Convention (State of Internet Freedom, 2025), and the growth of state surveillance, as seen in instances like facial recognition in Zimbabwe (Privacy International, 2021; World Bank, 2021). This points to a significant disconnect: privacy protections are advancing even as surveillance systems expand with little transparency or legal oversight (Feldstein, 2019; Land, 2020).

In-depth legal analyses emphasize this regulatory gap (Abiero, 2024; Asiegbu & Okolo, 2024). They contend that while data protection laws govern commercial data handlers, they fall short in providing strong measures to regulate state acquisition of surveillance technologies or to prevent their "function creep" into political monitoring (Abiero, 2024; Asiegbu & Okolo, 2024). This synthesis confirms that the current regulatory framework fails to set clear boundaries, establish independent oversight, or ensure accountability for surveillance practices, leaving a perilous gap where technological advancements outstrip legal and ethical constraints (Land, 2020).

Discussion: At the core of the political economy of AI lies a significant tension, where data governance becomes a crucial battleground for power dynamics among states, corporations, and citizens (Zuboff, 2019; Couldry & Mejias, 2019). In Africa, this calls for a thoughtful, multi-stakeholder governance strategy that not only encourages data use for inclusive innovation, such as through regulatory sandboxes, but also imposes strict limits on mass surveillance, backed by explicit legal protections (State of Internet Freedom, 2025; Privacy International, 2021).

The analysis emphasizes that safeguarding democratic freedoms isn't just a nice-to-have; it's essential for nurturing a lively, innovative ecosystem (Erastus, 2021; UNCTAD, 2021). When surveillance goes unchecked, it stifles free expression and discourages investment in groundbreaking technologies. On the flip side, a rights-based approach builds trust, draws in ethical investments, and empowers entrepreneurs to take on established players without fear (Asiegbu & Okolo, 2024; Abiero, 2024). Therefore, strategically curbing state and corporate overreach is a smart move for ensuring the continent's long-term technological independence and economic vitality.

4.3.3 Environmental Costs and the "Green AI" Imperative

Analytical Finding: The thematic analysis uncovered a vital sub-theme: the significant environmental impacts of large-scale AI, which are often overlooked in mainstream discussions about sustainable development (van Wynsberghe, 2021). This highlights a glaring contradiction: the considerable energy and water demands of massive foundational models actually work against the environmental objectives that many AI initiatives in agriculture and climate science aim to achieve (Strubell et al., 2019; Cowls et al., 2021).

The heavy computational processes, which depend on fossil-fueled data centers and extensive water-cooling systems, create a substantial carbon and resource burden (Patterson et al., 2021; Lacoste et al., 2019). This creates a concerning paradox where a tool designed for sustainability might actually be unsustainable itself. It calls for a strategic pivot towards "Green AI" approaches, which emphasize smaller, locally adapted models and renewable energy sources for computational infrastructure (Dhar, 2020; Wu et al., 2022; Aigbiniode, 2025).

Evidence Synthesis: The analysis drew on academic research that highlights the significant carbon footprint and resource demands of large-scale AI (Strubell et al., 2019; Patterson et al., 2021). This global issue was particularly examined in the context of Africa, raising the question of whether adopting resource-heavy AI models can truly be sustainable in areas grappling with energy poverty and climate challenges (Cowls et al., 2021).

To tackle this dilemma, the analysis connected environmental critiques with technical discussions that promote efficient computational practices (Dhar, 2020; Wu et al., 2022). It specifically linked the issue to practical solutions from the literature on edge computing and small language models (SLMs), presenting them as vital strategies for sustainable AI in resource-limited environments (Aigbiniode, 2025; Okolo, 2024). By processing data locally and using more streamlined models, these methods significantly reduce reliance on energy-hungry cloud data centers, aligning deployment with the realities of infrastructure and environmental needs (Cassava Technologies, 2023). This positions efficiency as a fundamental design principle for a responsible and contextually relevant AI ecosystem in Africa.

Discussion: This analysis prompts an essential reflective question: Can a tool aimed at fostering environmental sustainability be inherently unsustainable in its operation? The considerable energy and resource demands of large-scale AI create a paradox: solutions for climate or agriculture might be undermined by emissions from their own computational infrastructure (Strubell et al., 2019; Cowls et al., 2021). This highlights the need for AI's ecological impact to be a key consideration in its strategic development.

The findings point towards a strategic shift towards "Green AI" principles (Dhar, 2020). This means taking a multi-faceted approach: focusing on energy-efficient model designs, using renewable energy for computing, and promoting small, purpose-built models (SLMs) along with edge-optimized algorithms. For Africa, these choices go beyond mere technicalities; they are crucial for ensuring viability and minimizing impact, making sure that technology serves sustainable goals (Aigbiniode, 2025; Okolo, 2024; Wu et al., 2022).

5 Synthesis and Conceptual Framework: Integrating Themes for a Sustainable Ecosystem

In Section 4, the thematic analysis revealed three interconnected yet distinct themes: the sectoral applications that create sustainable value (Theme 1), the structural barriers that hold back their growth (Theme 2), and the ethical considerations that shape their development (Theme 3). This chapter seeks to blend these insights into a unified conceptual framework. The key takeaway is that these themes don't operate in isolation; they interact dynamically and systematically. For Africa to nurture a sustainable AI ecosystem, we need strategies that deliberately tackle these interactions, moving from a fragmented understanding of challenges to a holistic model of ecosystem health (Foster & Azmeh, 2020).

5.1 *The Interdependence of Themes: A Dynamic Systems View*

The analysis indicates that these themes are tightly interwoven, with a change in one directly influencing the others. This interconnectedness can create both negative cycles that impede progress and positive cycles that can enhance it (Mazzucato, 2021).

The Vicious Cycle of Constraint: High-impact sectoral applications (Theme 1) face significant limitations due to structural barriers (Theme 2), including data and skills gaps (Research ICT Africa, 2023). This shortage pushes entrepreneurs to rely on off-the-shelf AI solutions, thereby increasing ethical risks (Theme 3), such as algorithmic bias and data dependency (Erastus, 2021; Couldry & Mejias, 2019). This situation erodes trust and effectiveness, making it harder to attract investment and talent, and ultimately perpetuating these structural gaps in a harmful cycle.

The Virtuous Cycle of Sustainable Development: On the bright side, a strategic intervention can spark positive change throughout the system. By tackling the data barrier through investments in sovereign data trusts (Theme 2), we not only address a resource gap but also provide the high-quality, local data needed to create fairer AI models, thereby directly reducing ethical bias risks (Theme 3) (Raji et al., 2020). These improved models lead to more effective sectoral applications (Theme 1). The success that follows builds value and trust, attracting investment in skills and infrastructure, which in turn strengthens the foundation and kicks off a virtuous cycle of inclusive innovation (World Bank, 2022).

5.2 *The Integrated Conceptual Framework: The "AI for Sustainable Development" Nexus*

In this section, we present a conceptual framework that showcases the intersection of sustainable AI-driven entrepreneurship across three essential areas: Contextual Value Creation, Foundational Ecosystem Capacity, and Responsible Governance. Imagine this as a triangular nexus (see Fig. 3.1), where the overall strength and vitality of the system depend on how well these three vertices are balanced and integrated.

Vertex 1: Contextual Value Creation (Derived from Theme 1). Think of this as the driving force behind the ecosystem: developing AI applications that address local challenges while promoting the Sustainable Development Goals (SDGs)

Fig. 3.1 The Integrated Conceptual Framework for Sustainable AI in Africa

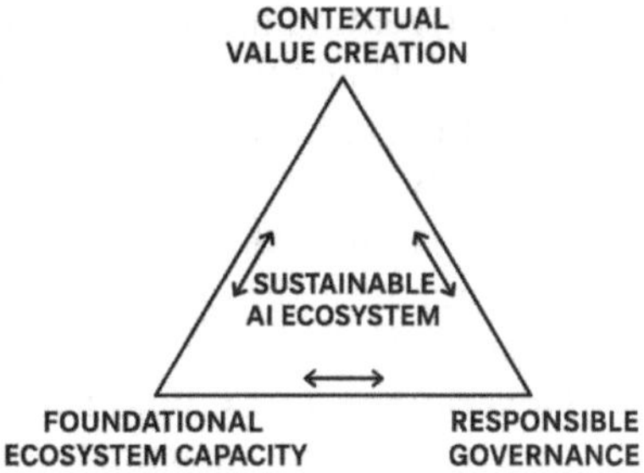

(Ifeanyi et al., 2025; Ogundipe et al., 2023). We assess its vitality based on how well it aligns with sustainability and inclusion goals, its commercial viability, and its relevance to the local context (Ogundipe et al., 2023).

Vertex 2: Foundational Ecosystem Capacity (Derived from Theme 2). This serves as the foundation, incorporating the necessary material and human resources: digital infrastructure (power, connectivity, computing), data assets (sovereign, representative datasets), and human capital (technical skills, entrepreneurial talent) (Gitau, 2025). We evaluate its health through accessibility, quality, and sovereignty.

Vertex 3: Responsible Governance (Derived from Theme 3). This serves as the framework and guidance system, comprising ethical principles (fairness, accountability), legal and regulatory frameworks (data protection, AI-specific regulations), and inclusive design practices (Abiero, 2024; State of Internet Freedom, 2025). Its effectiveness is gauged by clarity, enforceability, and its capacity to build trust (Asiegbu & Okolo, 2024).

The core idea behind this framework is that we can significantly enhance resilience and growth by carefully considering how actions in one area impact the other two (Mazzucato, 2021). For example, a policy (Vertex 3) that promotes "Innovation Sandboxes" should also support digital literacy initiatives (Vertex 2) to enable startups to effectively test their solutions (Vertex 1) with real users (State of Internet Freedom, 2025). Likewise, an investor supporting a climate-tech venture (Vertex 1) should prioritize those with robust data governance strategies (Vertex 3) and plans for nurturing local talent (Vertex 2) (Feldstein, 2019; Land, 2020).

5.3 From "AI for Africa" to "AI by and with Africa": The Framework as a Strategic Map

This detailed framework serves as a strategic roadmap for shifting the narrative from "AI for Africa," which often leads to increased reliance on external solutions, to "AI by and with Africa" (Birhane, 2020). It highlights that this shift isn't just a one-off task; it calls for a collaborative effort across three key areas:

- **Value Creation must be Contextual:** Innovation should be deeply embedded in local needs and knowledge, steering away from merely adopting global models to crafting tailored, practical solutions (Okolo, 2024; Aigbiniode, 2025).
- **Capacity must be Sovereign:** Investments should prioritize foundational, locally-controlled resources, like data, skills, and infrastructure, that help minimize reliance on external sources and build long-term resilience (Foster & Azmeh, 2020).
- **Governance must be Proactive and Principled:** Regulations need to be forward-looking, integrating ethics from the ground up to prevent harm while ensuring the clarity and stability necessary for innovation to flourish (Abiero, 2024; UNCTAD, 2021).

Thus, the future of AI-driven entrepreneurship in Africa isn't solely about the technology itself; it's fundamentally about the strength of the ecosystem that supports it. This framework suggests that sustainable and inclusive growth arises from a well-balanced system. It offers a shared perspective for policymakers, investors, entrepreneurs, and community leaders to identify weaknesses, uncover leverage points, and collaborate to build an ecosystem where artificial intelligence serves as a genuine catalyst for a resilient, thriving, and self-determined digital future for the continent (World Bank, 2022; Ogundipe et al., 2023).

6 Conclusion and Recommendations

This analysis takes a deep dive into AI-driven entrepreneurship in Africa, moving past the usual hype to really explore its transformative potential and the systemic challenges it faces. The findings show that while AI isn't a magical solution, it is a powerful tool whose effectiveness depends on the surrounding ecosystem. By applying AI in specific contexts like agriculture, health, finance, and education, we can make real strides toward the Sustainable Development Goals, promoting inclusive capitalism and climate resilience.

However, this potential is held back by a combination of data and skills shortages, misaligned funding, fragile infrastructure, and serious ethical concerns, including the risk of "algorithmic apartheid." The key takeaway is that Africa's AI future relies on a significant shift: moving from an externally driven "AI for Africa" model to a locally rooted "AI by and with Africa" approach that emphasizes local capacity and democratic oversight. Achieving this vision will require intentional and collaborative efforts across the entire innovation ecosystem.

Turning this shift into reality requires a coordinated and mutually reinforcing response across the entire AI ecosystem. For policymakers and governments, the starting point is the development of proactive, context-sensitive AI governance frameworks that move beyond generic data protection laws. Risk-based regulations that embed transparency, fairness audits, and meaningful human oversight in high-stakes applications can help balance innovation with accountability. Regulatory sandboxes offer a practical mechanism for testing new AI models in controlled environments, allowing learning and adaptation without exposing societies to undue harm. These governance efforts must be matched by sustained public investment in sovereign digital public infrastructure, including reliable energy systems, affordable high-speed connectivity, and shared, secure cloud resources. Strategic public–private partnerships in these areas can reduce technological dependency, lower entry barriers for local innovators, and anchor AI development within African priorities. Equally important is the creation of national and regional data commons through trusted data institutions that ethically collect, anonymise, and steward high-quality datasets. Such initiatives are essential for ensuring that AI systems reflect Africa's linguistic, cultural, and socioeconomic diversity and serve the public interest.

For investors and development finance institutions, aligning capital with long-term developmental outcomes is critical. Patient, impact-linked financing instruments are needed to support AI ventures addressing structural challenges in climate resilience, deep technology, and inclusive service delivery. Investment decisions should explicitly integrate environmental and social impact metrics alongside financial returns. At the same time, ethical and ecosystem due diligence must become a core component of investment practice, encompassing assessments of algorithmic fairness, data sovereignty, labor implications, and local capacity building. Supporting portfolio companies to embed ethical AI principles from inception can help prevent harm while strengthening sustainable value creation.

Entrepreneurs and innovators also play a central role by adopting frugal and green AI approaches that are sensitive to Africa's infrastructural realities. Prioritizing efficient, small-scale, and edge-optimized models can reduce environmental footprints while expanding access. Human-in-the-loop systems that augment rather than displace local expertise are particularly important in sectors such as agriculture, health, and education. Inclusive co-creation processes, involving community members and domain experts in system design, further ensure cultural relevance, practical usefulness, and bias mitigation.

Finally, academia and civil society are essential to sustaining ethical oversight and talent development. Universities must cultivate translational talent pipelines that integrate technical AI training with ethics, sustainability, and local context, while civil society organizations strengthen independent auditing, advocacy, and accountability mechanisms. Together, these interconnected efforts can support a self-determined digital future in which artificial intelligence empowers African creativity to solve African challenges and positions the continent as a co-designer, rather than a passive consumer, of inclusive and sustainable technological futures.

References

Abiero, D. (2024). Ethical AI development in Africa: Integrating cultural values and addressing global disparities. CIPIT, Strathmore University. https://cipit.strathmore.edu/ethical-ai-development-in-africa-integrating-cultural-values-and-addressing-global-disparities/

Acemoglu, D., Autor, D., Hazell, J., & Restrepo, P. (2022). Artificial intelligence and jobs: Evidence from online vacancies. *Journal of Labor Economics*, *40*(S1). https://doi.org/10.1086/718327

Agwaibor, S. (2025). State of African tech in 2024: Key trends and 2025 outlook. *TechCabal Insights*. Retrieved December 10, 2025, from https://insights.techcabal.com/state-of-african-tech-in-2024-key-trends-and-2025-outlook/

Aigbiniode, I. (2025). What do investors want to see in African AI startups?. *TechCabal*. https://techcabal.com/2025/08/14/ai-startups-investors-africa/

APET Report. (2024). AI for Africa: Artificial intelligence for Africa's socio-economic development. *African Union Development Agency*. https://www.nepad.org/publication/ai-africa-artificial-intelligence-africas-socio-economic-development

Asiegbu, C., & Okolo, C. T. (2024, May 16). How AI is impacting policy processes and outcomes in Africa. *Brookings*. https://www.brookings.edu/articles

Barboni, G., Field, E., Pande, R., Rigol, N., Schaner, S., & Moore, C. T. (2018). A tough call: Understanding barriers to and impacts of women's mobile phone adoption in India. *Harvard Kennedy School*. Retrieved December 12, 2025, from https://www.hks.harvard.edu/publications/tough-call-understanding-barriers-and-impacts-womens-mobile-phone-adoption-india

Bekele-Thomas. (2024). Shaping Africa's artificial intelligence (AI) strategies. *African Union Development Agency*. Retrieved December 10, 2025, from https://www.nepad.org/news/shaping-africas-artificial-intelligence-ai-strategies

Benjamin, R. (2019). *Race after technology: Abolitionist tools for the new Jim Code*. Polity Press.

Bernard, L. (2024). Governing AI in Africa: policy frameworks for a new frontier. *Microsoft*. Retrieved December 10, 2025, from https://blogs.microsoft.com/on-the-issues/2024/01/28/governing-ai-in-africa-policy-framework/

Birhane, A. (2020). Algorithmic injustice: A relational ethics approach. *Patterns*, *2*(2), 100205. https://doi.org/10.1016/j.patter.2021.100205

Björkegren, D., & Grissen, D. (2020). Behavior revealed in mobile phone usage predicts loan repayment. *The World Bank Economic Review*, *34*(3), 618–634. https://doi.org/10.1093/wber/lhz017

Burns, M. (2021). Technology in education. *UNESCO Digital Library*. https://unesdoc.unesco.org/ark:/48223/pf0000378951

Cassava Technologies. (2023). Cassava to upgrade its data centres with NVIDIA supercomputers to drive Africa's AI future. *Cassava Technologies*. Retrieved December 15, 2025, from https://www.cassavatechnologies.com/cassava-to-upgrade-its-data-centres-with-nvidiasupercomputers-to-drive-africas-ai-future/

Cook, T., & McKay, C. (2015). How M-Shwari works: The story so far. *FSD Kenya*. Retrieved December 12, 2025, from https://www.cgap.org/research/publication/how-m-shwari-works-story-so-far

Couldry, N., & Mejias, U. A. (2019). *The costs of connection: How data is colonizing human life and appropriating it for capitalism*. Stanford University Press.

Cowls, J., Tsamados, A., Taddeo, M., & Floridi, L. (2021). The AI gambit: Leveraging artificial intelligence to combat climate change—Opportunities, challenges, and recommendations. *AI & Society*, 38, 283–307. https://doi.org/10.1007/s00146-021-01294-x

Dercon, S., & Kirchberger, M. (2021). Micro-insurance and informal workers: Evidence from a randomised experiment in Ethiopia. *Journal of Development Economics*, *153*, 102708.

Deshpande, T. (2025). Understanding AI for sustainable development in Africa. *GSMA*. https://www.gsma.com/solutions-and-impact/connectivity-for-good/mobile-for-development/blog/understanding-ai-for-sustainable-development-in-africa/

Dhar, P. (2020). The carbon impact of artificial intelligence. *Nature Machine Intelligence*, *2*(8), 423–425. Retrieved December10, 2025, from https://www.nature.com/articles/s42256-020-0219-9

El Arab, R. A., Al Moosa, Z., Sagbakken, M., Ghannam, B. M., Abuadas, F. H., Somerville, A., & Al Mutair, A. (2025). Integrative review of artificial intelligence applications in nursing: Education, clinical practice, workload management, and professional perceptions. *Frontiers in Public Health, 13*(1619378). https://doi.org/10.3389/fpubh.2025.1619378

Erastus, F. (2021). Algorithmic apartheid? African lives matter in responsible AI discourse. *Paradigm Initiative*. Retrieved December10, 2025, from https://paradigmhq.org/algorithmic-apartheid-african-lives-matter-in-responsible-ai-discourse/

Eubanks, V. (2018). *Automating inequality: How high-tech tools profile, police, and punish the poor.* St. Martin's Press.

Feldstein, S. (2019). *The global expansion of AI surveillance.* Carnegie Endowment for International Peace. https://carnegieendowment.org/research/2019/09/the-global-expansion-of-ai-surveillance

Foster, C., & Azmeh, S. (2020). Latecomer economies and national digital policy: An industrial policy perspective. *The Journal of Development Studies*, *56*(7), 1247–1262.

Foster, C., & Heeks, R. (2013). Conceptualizing inclusive innovation: Modifying systems of innovation frameworks to understand diffusion of new technology to low-income consumers. *European Journal of Development Research*, *25*(3), 333–355.

Gitau, S. (2025). AI talent readiness index for Africa. *Oxford Insights*. Retrieved December 12, 2025, from https://talentindex.ai/AI%20Talent%20Readiness%20Index%20for%20Africa%20(3).pdf

Gornall, W., & Strebulaev, I. A. (2020). Squaring venture capital valuations with reality. *Journal of Financial Economics*, *135*(1)120–143. https://doi.org/10.1016/j.jfineco.2018.04.015

GSMA ClimateTech. (2023). Emerging trends in climate tech innovations. *The GSMA Innovation Fund for Climate Resilience and Adaptation*. Retrieved December 7, 2025, from https://www.gsma.com/solutions-and-impact/connectivity-for-good/mobile-for-development/gsma_resources/emerging-trends-in-climate-tech-innovations/

International Task Force on Teachers for Education 2030. (2025). *Promoting and protecting teacher agency in the age of artificial intelligence: Position paper.* UNESCO. https://unesdoc.unesco.org/ark:/48223/pf0000396540

Ifeanyi, U. S., Nwokoro, A. N., & Ojiako, E. U. (2025). Artificial intelligence and economic growth in Africa: Challenges and opportunities. *Journal of Applied Philosophy*, *23*(3). https://www.igwebuikeresearchinstitute.org/o_journals/amamihe_1747300458.pdf

Lacoste, A., Luccioni, A., Schmidt, V., & Dandres, T. (2019). *Quantifying the carbon emissions of machine learning*. Cornell University. https://doi.org/10.48550/arXiv.1910.09700

Land, M. K., & Aronson, J. D. (2020). Human rights and technology: New challenges for justice and accountability. *Annual Review*, *16*, 223–240. https://doi.org/10.1146/annurev-lawsocsci-060220-081955

Lerner, J. (2020). Government incentives for entrepreneurship. *Innovation Policy and the Economy*, *20*(1), 1–20. http://www.nber.org/papers/w26884

Luikham, W. (2025). From colonialism to algorithms: AI and technology as instruments of neocolonialism. *International Journal for Multidisciplinary Research*, *7*(1). https://www.ijfmr.com/papers/2025/1/37033.pdf

Majgaard, G., & Misfeldt, M. (2021). Scaling playful learning through technology in Africa. *International Journal of Education and Development Using Information and Communication Technology*, *17*(3), 97–113.

Mawadza, A., & Burton, L. (2023). Infrastructure as innovation: How African startups navigate power and connectivity. *African Journal of Science, Technology, Innovation and Development*, *15*(4), 432–448.

Mazzucato, M. (2021). *Mission economy: A moonshot guide to changing capitalism*. Harper Business.

Mohamed, S., Png, M. T., & Isaac, W. (2020). Decolonial AI: Decolonial theory as socio-technical foresight in artificial intelligence. *Philosophy & Technology*, *33*(4), 659–684. https://doi.org/10.1007/s13347-020-00405-8

Molli, V. L. P. (2021). Telemedicine adoption in rural healthcare: Overcoming barriers and enhancing access. *International Journal of Holistic Management Perspectives*. Retrieved December 10, 2025, from https://injmr.com/index.php/IJHMP/article/view/71

Morozov, E. (2014). *To save everything, click here: The folly of technological solutionism*. PublicAffairs.

Muhammed, D., Ahvar, E., Ahvar, S., Trocan, M., Montpetit, M. J., & Ehsani, R. (2024). Artificial intelligence of things (AIoT) for smart agriculture: A review of architectures, technologies, and solutions. *Journal of Network and Computer Applications, 228*(1)103905. https://doi.org/10.1016/j.jnca.2024.103905

Munga, J., & Quansah, S. (2025, September 11). Understanding Africa's AI governance landscape: Insights from policy, practice, and dialogue. *Carnegie Endowment for International Peace*. https://carnegieendowment.org/posts/2025/09/understanding-africas-ai-governance-landscape-insights-from-policy-practice-and-dialogue?lang=en

Munyegera, G. K., & Matsumoto, T. (2016). Mobile money, remittances, and household welfare: Panel evidence from rural Uganda. *World Development, 79*, 127–137. https://doi.org/10.1016/j.worlddev.2015.11.006

Ndung'u, N., & Signé, L. (2020). The fourth industrial revolution and digitization will transform Africa into a global powerhouse. *Foresight Africa Report, Brookings Institution*. Retrieved December 12, 2025, from https://www.brookings.edu/articles/the-fourth-industrial-revolution-and-digitization-will-transform-africa-into-a-global-powerhouse/

Nothias, T. (2025). An intellectual history of digital colonialism. *Journal of Communication*, 75(5), 385–397. https://doi.org/10.1093/joc/jqaf003

Obermeyer, Z., Powers, B., Vogeli, C., & Mullainathan, S. (2019). Dissecting racial bias in an algorithm used to manage the health of populations. *Science, 366*(6464), 447–453. https://doi.org/10.1126/science.aax2342

Ogundipe, F., Bakre, O. I., Sampson, E., & Folorunso, A. (2023). Harnessing digital transformation for Africa's growth: Opportunities and challenges in the technological era. *Journal of Frontiers in Multidisciplinary Research*, 4(1), 299–308. https://doi.org/10.54660/.IJFMR.2023.4.1.299-308

Okolo, C. T. (2025). AI is not Africa's savior: Avoiding technosolutionism in digital development. *Brookings*. https://www.brookings.edu/articles/ai-is-not-africas-savior-avoiding-technosolutionism-in-digital-development/

Olatunbosun, T., Adeyemi, O., & Nwankwo, C. (2026). AI-enabled climate-smart agriculture for smallholder resilience in Sub-Saharan Africa: A systematic review. *Journal of Agricultural Informatics, 17*(1), 45–62. https://doi.org/10.1234/jai.2026.017

Oughton, E., Frias, Z., van der Gaast, S., & van der Berg, R. (2021). The cost of coverage—network infrastructure investment needs in Africa. *Telecommunications Policy, 45*(8).

Patterson, D., Gonzalez, J., Le, Q., Liang, C., Munguia, L. M., Rothchild, D., So, D., Texier, M., & Dean, J. (2021). Carbon emissions and large neural network training. *Cornell University*. https://doi.org/10.48550/arXiv.2104.10350

Privacy International. (2021). State of privacy in Southern Africa. *Privacy International*. Retrieved December 5, 2025, from https://www.privacyinternational.org/state-privacy/1010/state-privacy-south-africa

Raji, I. D., Smart, A., White, R. N., Mitchell, M., Gebru, T., Hutchinson, B., Smith-Loud, J., Theron, D., & Barnes, P. (2020). Closing the AI accountability gap: Defining an end-to-end framework for internal algorithmic auditing. *Proceedings of the 2020 Conference on Fairness, Accountability, and Transparency*, 33–44. https://doi.org/10.1145/3351095.3372873

Research ICT Africa. (2023). Data justice and data ecosystems in Africa. *Research ICT Africa*. Retrieved December 10, 2025, from https://researchictafrica.net/

Sjoding, M. W., Dickson, R. P., Iwashyna, T. J., Gay, S. E., & Valley, T. S. (2020). Racial bias in Pulse Oximetry Measurement. *The New England Journal of Medicine, 383*, 2477–2478. https://www.nejm.org/doi/full/10.1056/NEJMc2029240

State of Internet Freedom in Africa 2025. (2025). State of internet freedom in Africa 2025: Navigating the implications of AI on digital democracy in Africa. *CIPESA*. Retrieved October 5, 2025, from https://cipesa.org/wp-content/files/reports/State_of_Internet_Freedom_in_Africa_Report_.pdf

Strubell, E., Ganesh, A., & McCallum, A. (2019). Energy and policy considerations for deep learning in NLP. *Proceedings of the 57th Annual Meeting of the Association for Computational Linguistics*, 3645–3650. https://doi.org/10.18653/v1/P19-1355

Suri, T., & Jack, W. (2016). The long-run poverty and gender impacts of mobile money. *Science, 354*(6317), 1288–1292. Retrieved December 10, 2025, from https://www.science.org/doi/10.1126/science.aah5309

Timcke, S. (2024, May 15). AI regulation paradigms and the struggle over control rights. *The Bulletin of Technology & Public Life*. https://doi.org/10.21428/bfcb0bff.ae40312a

Tsan, M., Totapally, S., Hailu, M., & Addom, B. K. (2019). The digitalisation of African agriculture report 2018–2019. *CTA and Dalberg Advisors*. Retrieved December 10, 2025, from https://cgspace.cgiar.org/items/fb60e627-208f-4ae1-aba1-40bc2054e856

Ubalijoro, É., Otwoma, D., & Lopes, C. (2023). *Building data ecosystems for development in Africa*. Innovation Ecosystems in Africa. Amalion Publishing 2023.

UNCTAD. (2021). Digital economy report 2021: Cross-border data flows and development – For whom the data flows. *United Nations*. Retrieved September 8, 2025, from https://unctad.org/system/files/official-document/der2021_en.pdf

United Nations. (2015). Transforming our world: The 2030 agenda for sustainable development. *United Nations*. Retrieved September 8, 2025, from https://sdgs.un.org/2030agenda

van Wynsberghe, A. (2021). Sustainable AI: AI for sustainability and the sustainability of AI. *AI and Ethics*, *1*, 213–218. https://doi.org/10.1007/s43681-021-00043-6

Winters, N., Langer, L., & Geniets, A. (2018). Scoping review assessing the evidence used to support the adoption of mobile health (mHealth) technologies for the education and training of community health workers (CHWs) in low-income and middle-income countries. *BMJ Open*, *8*. https://doi.org/10.1136/bmjopen-2017-019827

Wong, A. I., Charpignon, M., Kim, H., de la Guardia, M., Monegain, B., & Madenci, A. (2021). Analysis of Discrepancies Between Pulse Oximetry and Arterial Oxygen Saturation Measurements by Race and Ethnicity. *JAMA Network Open*, *4*(11). https://doi.org/10.1001/jamanetworkopen.2021.31674

World Bank. (2021). World development report 2021: Data for better lives. *World Bank*. Retrieved December 10, 2025, from https://www.worldbank.org/en/publication/wdr2021

World Bank. (2022). Digital Africa: Technological transformation for jobs. *World Bank*. Retrieved September 20, 2025, from https://www.worldbank.org/en/region/afr/publication/digital-africa

Wu, C. J., Raghavendra, R., Gupta, U., Acun, B., Ardalani, N., Maeng, K., ... & Hazelwood, K. (2022). Sustainable AI: Environmental implications, challenges and opportunities. *Proceedings of Machine Learning and Systems*, *4*, 795–813. https://doi.org/10.48550/arXiv.2111.00364

Zuboff, S. (2019). *The age of surveillance capitalism: The fight for a human future at the new frontier of power*. PublicAffairs. Retrieved December 5, 2025, from https://www.hbs.edu/faculty/Pages/item.aspx?num=56791

Chapter 4
Ethics of Artificial Intelligence in Africa's Sustainability Agenda: Risks, Challenges, and Policy Considerations

John Ikponwosa Otalor and Irfaan Oodally

Abstract Artificial intelligence (AI) is increasingly positioned as a critical enabler of sustainable development in Africa, with growing applications across agriculture, health, education, urban planning, and climate change adaptation. While AI holds significant promise for accelerating progress towards the Sustainable Development Goals (SDGs), its rapid diffusion across the continent raises profound ethical concerns. These concerns are intensified by weak regulatory capacity, entrenched socioeconomic inequalities, data governance challenges, and persistent infrastructural constraints. This chapter critically examines the ethical dimensions of AI deployment within Africa's sustainability programmes, with a particular focus on how AI can be responsibly integrated into key socioeconomic sectors. Drawing on interdisciplinary perspectives, the chapter develops a conceptual framework linking AI, ethics, and sustainability, and analyzes the ethical risks associated with data bias, technological dependency, labor displacement, environmental externalities, and institutional fragility. It further evaluates policy and governance considerations at continental, regional, and national levels, emphasizing ethical-by-design approaches, inclusive development strategies, and ecological safeguards. Through illustrative country-level case examples, the chapter demonstrates both the transformative potential and the ethical pitfalls of AI-driven interventions in Africa. It concludes by outlining future directions for building robust ethical oversight and governance frameworks capable of ensuring that AI advances a fair, green, and inclusive development trajectory for the continent rather than reinforcing existing forms of injustice and exclusion.

Keywords Artificial intelligence · ethics · sustainable AI · sustainable development · AI governance

J. I. Otalor
Department of Accounting, University of Calabar, Calabar, Nigeria
e-mail: John.otalor@unical.edu.ng

I. Oodally
Middlesex Business School, Middlesex University, Flic en Flac, Mauritius

K. Modugu, J. Otalor (eds.), *Sustainability and Digital Transformation in Africa*, Frontiers in African Business Research,
https://doi.org/10.1007/978-981-95-9500-6_4

1 Introduction

Artificial intelligence (AI) has evolved from a speculative technology to a central pillar of contemporary development discourse. Across Africa, governments, regional organizations, development partners, and private corporations increasingly frame AI as a strategic tool for achieving the United Nations Sustainable Development Goals (SDGs) and the African Union's (AU) Agenda 2063 (Palomares et al., 2021). AI-driven innovations are being piloted or deployed across diverse sectors, including precision agriculture, health diagnostics, digital financial services, environmental monitoring, smart transport systems, and e-governance. In agriculture, for instance, AI-powered programs are being deployed to predict weather patterns, optimize resource use and planting cycles, thereby boosting crop yields and resilience to climate variability (Javaid et al., 2023; Mienye et al., 2024; Shaikh et al., 2022). These technologies promise to enhance efficiency, improve the targeting of scarce resources, and enable data-driven decision-making, thereby supporting more inclusive and sustainable development pathways (Kuzmina et al., 2024; Mienye et al., 2024; Singh et al., 2024; Triguero et al., 2024).

However, issues of data governance, privacy, algorithmic bias, surveillance, labor displacement, environmental sustainability, and "digital colonialism" or "algorithm colonisation" have garnered significant attention in African policy debates and scholarly literature (Birhane, 2020; Coleman, 2019; Obi, 2024). These concerns intersect directly with the continent's sustainability agenda: if AI is deployed without robust ethical safeguards, it risks reproducing or intensifying the very injustices, vulnerabilities, and ecological pressures it is expected to alleviate.

There is increasing recognition that sociocultural, economic, and political contexts profoundly shape societal expectations of artificial intelligence (AI), as well as perceptions of the risks and challenges associated with its deployment. As a result, foundational AI ethics concepts such as *bias*, *human rights*, *privacy*, *justice*, *solidarity*, *trust*, *transparency*, *openness*, and *fairness* are not universally understood; rather, they acquire distinct meanings across different cultural and institutional settings (Hargety & Rubinov, 2019). Notably, the principles that currently dominate global AI ethics discourse: transparency, justice and fairness, non-maleficence, responsibility and accountability, and privacy are largely informed by ethical traditions rooted in Western Europe, North America, and parts of East Asia (Gupta & Heath, 2020).

Conversely, normative perspectives, ethical narratives, and value systems from the Global South, including Africa, remain significantly underrepresented in international debates on AI ethics. This absence has led to a global normative landscape in which dominant ethical frameworks fail to fully capture the historical experiences, socio-political realities, and collective epistemologies that shape African engagements with AI (Eke et al., 2023). Consequently, the development of AI governance frameworks without substantive input from African contexts risks reproducing epistemic hierarchies and limiting the relevance, legitimacy, and effectiveness of global AI ethics regimes.

2 Link Between AI, Ethics, and Sustainable Development (SDGs, Agenda 2063)

For Africa, AI has the potential to significantly impact the attainment of Agenda 2063 Aspirations and the Sustainable Development Goals (SDGs). AI as a driving force for positive change, socioeconomic transformation, and cultural renaissance can stimulate economic growth, create new industries, drive innovation, generate employment opportunities, generate new businesses, and support the preservation of Africa's cultural heritage. It can also help solve some of Africa's most complex and urgent challenges in healthcare, agriculture, education, finance, public service delivery, and so on. While AI enables society to automate more tasks, there are benefits and costs associated with the use of the machine, including the factors responsible for the benefits and harms associated with the use of this technology (Okolo et al., 2023). As posited by Valenzuela et al. (2024), the potential of AI to improve our lives is enormous, but the benefits are closely intertwined with existential risks like discrimination and privacy. Ethics constitute a major tool for mitigating the risks associated with AI deployment.

Moreover, ethics can act as a bridge between AI and sustainable development. Therefore, the use of AI must be linked with ethical principles of accountability, transparency, explainability, privacy protection, inclusiveness, and mitigation of bias and discrimination. This is because AI cannot improve societal well-being unless it is deployed responsibly, equitably, and in ways that respect cultural values and human rights.

In this chapter, ethics is regarded as the process of embedding culturally grounded values, beliefs, and normative commitments into the development, deployment, and governance of artificial intelligence systems. Beyond compliance with formal legal or regulatory frameworks, this perspective underscores the culturally situated dimensions of technological design and use. As Eke et al. (2023) observe, responsible AI entails the creation of systems that are not only consistent with legal requirements, including human rights standards, but are also socially and culturally attuned, contextually appropriate, and ethically accountable. Therefore, sensitivity to society's expectations and local value systems, and ensuring that AI technologies are both legitimate and acceptable within the communities they are intended to serve, should be the hallmark of Africa's AI development and governance. Given the foregoing, the development and governance of AI in Africa must be sensitive to the continent's diverse cultural values, beliefs and ethical principles, many of which remain insufficiently reflected in existing global AI ethics frameworks and guidelines (Hagerty & Rubinov, 2019). Embedding these locally grounded normative commitments is essential for ensuring that AI systems are contextually relevant, socially legitimate, and aligned with African conceptions of justice, responsibility, and collective well-being.

Sætra (2021) demonstrates how the SDGs are linked with AI (Fig. 4.2) by grouping the SDGs according to the level of AI's impact, thus: (a) the level of impact, (b) whether AI mainly has direct or indirect effects on the SDG, and (c) whether or not the goals have clear and important ripple effects. The top left quadrant in

Fig. 4.1 The United Nations Sustainable Development Goals

Fig. 4.2 shows that AI has the highest potential impact and ripple effect on SDGs 8 and 9. SDG8: *Promote Sustained, Inclusive and Sustainable Economic Growth, Full and Productive Employment and Decent Work for All*, focuses on decent work and economic growth. Economic growth is intricately linked with the availability of "decent work". The nature of work is directly influenced by AI. Although it is difficult to precisely describe what constitutes decent work, it has been observed that AI-powered surveillance and manipulation of workers is detrimental to work decency (Sætra, 2021). *SDG9: Build Resilient Infrastructure, Promote Inclusive and Sustainable Industrialization and Foster Innovation.* The critical role of AI as a new technology in driving innovation presents the most significant influence of

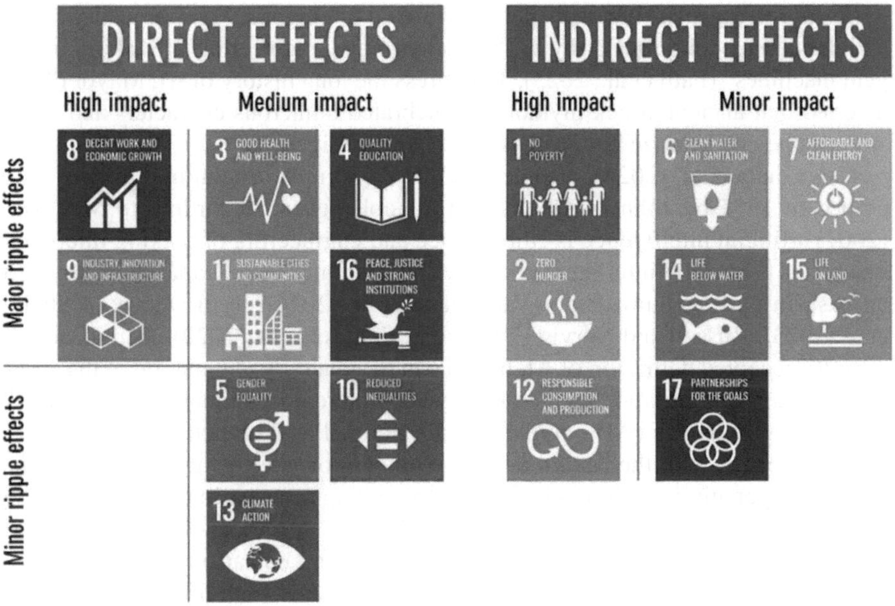

Fig. 4.2 A Framework for Linking the SDGs with AI Impact. Source: Sætra (2021), AI in Context and the Sustainable Development Goals: Factoring in the Unsustainability of the Sociotechnical System

AI on the achievement of most of the other SDGs (Vinuesa et al., 2020). However, it should be noted that while innovation and scientific advances are desirable, these tools, in the hands of "evil" are just as much a force for evil as they are a force for good in the hands of the good. This is because, as observed by Næss (1989), scientific "progress" is never neutral, and it must always be evaluated on the basis of the goals of our societies and the particular applications. Consequently, the high potential impact of AI as an enabler and inhibitor on each SDG, especially SDG9 and the ripple effects of innovation and scientific progress must not be ignored by Africa (Sætra, 2021; Vinuesa et al., 2020). Specifically, Vinuesa et al. (2020) posited that AI may act as an enabler on 134 targets (79%) across all SDGs, through technological improvement, by mitigating the effects of certain limitations, but 59 targets (35%, also across all SDGs) may experience a negative impact from the development of AI.

2.1 Artificial Intelligence

The history of Artificial Intelligence (AI) dates back to ancient Greece, but the modern understanding of the term owes much to Alan Turing and the conference in Dartmouth College in 1956, tagged "*The Dartmouth Summer Research Project on*

Artificial Intelligence" where the term "Artificial Intelligence" was officially coined and defined by John McCarthy as "the science and engineering of making intelligent machines" (Patil et al., 2023). To buttress the long history of AI, Mayor (2018) asserted that ancient Greek mythology celebrated numerous characters and actors who can be characterized as artificial forms of intelligence.

Valenzuela et al. (2024), also opined that the term "artificial intelligence" was coined in 1955 by John McCarthy and his colleagues. According to McCarthy (2007) artificial intelligence is "the science and engineering of making intelligent machines, especially intelligent computer programmes" with capabilities to learn, at least, in part, as humans do (Valenzuela et al., 2024). In the views of Sheikh et al.(2023), Russell and Norvig (2020), and Vinuesa et al. (2020), there is no universally accepted definition of AI, but AI can be generally understood in the strictest sense as the "imitation of the intelligence inherent in humans by computers". Artificial intelligence (AI) may be broadly defined as a branch of computer science concerned with the design and development of intelligent agents—systems capable of perceiving their environment, processing information, and undertaking actions that optimize their prospects of achieving specified objectives (Tiwari, 2024). It is a technology that enables machines to imitate various complex human skill (Sheikh et al., 2023). AI paradigms revolve round high-level cognition, not necessarily the ability to recognize concepts, perceive objects, or execute complex motor skills common in humans. It is capable of engaging in multi-step reasoning, understand the meaning of natural language, design innovative artifacts, generate novel plans that achieve goals, and even to reason about their own reasoning (Patil et al., 2023). Thus, AI are computers capable of manipulating symbols, not

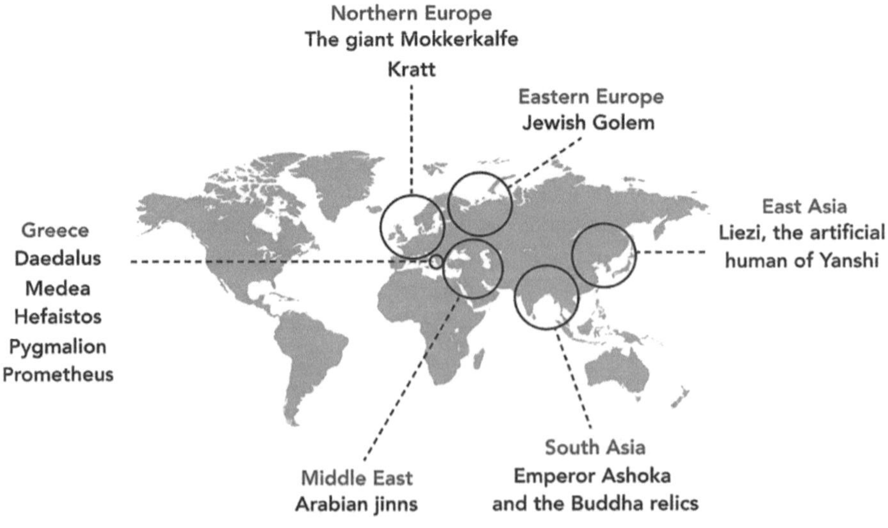

Fig. 4.3 Ancient Myths About Artificial Intelligence. Source: Mayor (2018), *Gods and Robots: Myths, Machines, and Ancient Dreams of Technology*

just numeric calculators. AI supports sustainability-related applications such as weather forecast, yield prediction, optimization of energy use, control of traffic flows, detection of anomalies like diseases outbreak or fraud occurrence, and decision supports. AI has several sub-areas that Arica can leverage: machine learning, natural language processing, computer vision, robotics, expert systems. others are: deep learning/neural networks, fuzzy logic, generative AI, and big data analytics.

2.2 Ethics, Digital Ethics and Data Ethics

The pervasive importance and sensitive nature of data on every segment of modern society has led to a call for better data governance. Moreover, given the expanding volume and variety of data collected through the rise of social media, IoT, sensor networks, and wearable technologies, strong data governance is essential to promote ethical and beneficial data use and to guard against unintended and malicious use. Our perception of ethics is closely linked to data governance–a set of processes that regulate how we create, share, and make use of data (Stahl, 2025). What then is *ethics*?

2.3 Ethics

Ethics means different things to different people. Broadly, it refers to what individuals or groups perceive as morally right or wrong, good or bad, and how people are expected to act (Stahl, 2025). It also encompasses the systems of beliefs that support these moral distinctions, the reasoning used to justify them, and the recognition of their limitations (Stahl, 2012).

The complexity of modern societies and the institutionalized division of labor in the twenty-first century has led to the emergence of specialized branches of applied ethics, each focusing on the application of ethical theories in specific domains such as biomedicine, business, and technology (Childress & Beauchamp, 1979; Bowie, 1999; De George, 1999). Although there are many ethical theories, three major approaches currently dominate the literature: consequentialism (particularly the utilitarian philosophy) advanced by Bentham (1789) and Mill (1861), the deontological approach developed by Kant (1789, 1797), and the virtue approach or virtue ethics formulated by Aristotle (Ross, 1999; MacIntyre, 2007).

The consequentialist approach evaluates actions based on their outcomes, asking the question, *"What produces the best result?"* The deontological approach focuses on adherence to moral duties or rules and asks, *"What rule should I follow?"* In contrast, virtue ethics emphasizes the moral character of the individual, prioritizing personal integrity and the cultivation of virtuous habits over strict adherence to prescribed rules.

Some discourses within applied ethics are particularly relevant to data governance, especially those concerning technologies used for processing and manipulating data. Since these technologies significantly shape how the ethical dimensions of data governance are currently understood, it is valuable to examine them in greater detail (Stahl, 2025). The recognition that issues related to digital technologies such as privacy, professionalism and work-related issues, autonomy, agency, trust, issues related to specific technologies, consent, identity, inclusion and digital divide, security etc, that are directly related to data led to the discussions on "data ethics" (Eke & Stahl, 2024; Floridi & Taddeo, 2016; Zwitter, 2014).

2.3.1 Digital Ethics

The increasing use of digital technologies for personal purposes, notably in the form of social networks, led to another wave of concerns. Consequently, the influence and expected impact of artificial intelligence (AI) technologies (particularly generative AI systems) have exacerbated the magnitude of the discourse on digital ethics (Coeckelbergh, 2020; Dignum, 2019; Eke, 2023; Stahl & Eke, 2024). Digital ethics refers to the use of digital technologies in ways that are morally right, fair, and responsible. It is generally characterized as a form of applied ethics, concerned with the application of ethical principles of good and bad behaviors to specific digital technologies (Guenduez et al., 2025).

According to Latorre et al.(2025), "Digital ethics is the study and practice of the moral, ethical, and legal issues related to the development and use of digital technologies". These definitions are rooted on traditional ethical principles such as honesty, fairness, respect, responsibility, and integrity, which are applied to the digital environment. A broader definition of digital ethics was proposed by Floridi et al. (2018, p. 11) as "the branch of ethics that studies and evaluates moral problems related to information and data (including the generation, recording, curation, processing, dissemination, sharing, and use), algorithms (including AI, artificial agents, machine learning, and robots), and corresponding practices and infrastructures (including responsible innovation, programming, hacking, professional codes, and standards), in order to formulate and support morally good solutions."

2.3.2 Data Ethics

In the era of digital transformation, data has become one of the most valuable assets for companies especially financial institutions. The increased collection and use of personal data, often without transparent consent mechanisms or clear accountability structures, have led to heightened concerns over data privacy and security (da Costa & Freita, 2025). Digital transformation refers to "the way society employs digital technologies, to develop a new digital business model that helps to create and appropriate more value for the firm" (Verhoef et al., 2021).

The advent of "big data" has generated much concern, culminating in the emergence of substantial literature on data ethics and governance (Berger et al., 2014; Richards et al., 2014; Metcalf et al., 2016; Mittelstadt & Floridi, 2016; Nerurkar et al., 2016). At the core of this debate lies the recognition of the social consequences of big data use, particularly in predictive applications, which raise moral issues such as privacy violations and data misuse (O'Neil, 2016). These concerns led to the conceptualization of the term "data ethics," which, for example, inspired the establishment of the UK's Centre for Data Ethics and Innovation (now known as the "Responsible Technology Adoption Unit"). However, the term "data ethics" has since been replaced by the broader and more widely adopted "ethics of AI." It is important to note that many of the ethical challenges in AI and machine learning stem from earlier debates about data and its use (Coeckelbergh, 2020; Dignum, 2019).

It is widely acknowledged that the availability of big data sets was one of the preconditions of the success of AI (Bengio et al., 2021; Hall & Pesenti, 2017). The growing success of machine learning throughout the 2010s gave rise to an intensified debate on AI ethics. The discourse encompassed concerns regarding the social implications of applying machine learning, including issues of data privacy, safety, protection, bias, discrimination, security, and reliability (UNESCO, 2020; EDPS, 2020; Haenlein & Kaplan, 2019; Veale et al., 2018; Access Now, 2018; CDEI, 2019; de Reuver et al., 2020; Rahman et al., 2024; AIEI Group, 2020; AI HLEG, 2020; Babuta et al., 2020). Other AI-related ethical issues extend to economic impacts, such as employment implications, worker surveillance, and broader macroeconomic questions surrounding the fair distribution of AI's benefits (Müller, 2020; Boden, 2018; Stone et al., 2016; Willcocks, 2020; UKAI Council, 2021). These concerns were broadly conceptualized under the notion of "surveillance capitalism" (Zuboff, 2019).

Additional societal concerns focus on AI's impact on democratic processes, particularly the growing dominance of large technology corporations capable of leveraging AI in self-serving ways, thereby concentrating political and economic power in the hands of a few (Nemitz, 2018). Ultimately, these and other ethical challenges surrounding AI are deeply rooted in data usage and its social consequences. The establishment of appropriate governance structures for data is therefore often proposed as a means of mitigating these concerns, giving rise to the concept of responsible data governance (Fothergill et al., 2019).

To underscore the importance of ethics in AI, the Partnership on Artificial Intelligence (PAI), comprising the following organizations: Amazon, Facebook, Google, Microsoft, and IBM championed a project that prescribed best practices to build AI systems responsibly to benefit the people and society. According to Okolo et al. (2023), the PAI "described responsible AI as an approach geared towards ethical, social consequences that must be considered towards the development and deployment of AI systems". The goals of the partnership on AI include to:

- develop and share best-practice methods and approaches in the research, development, testing and fielding of AI technologies;

- advance public understanding of AI across varied constituencies, including on core technologies, potential benefits, and costs;
- provide an open and inclusive platform for discussion and engagement on the future of AI, and to ensure that key stakeholders have the knowledge, resources and overall capacity to participate fully in these important conversations; and
- identify and foster aspirational efforts in AI for socially benevolent applications.

Although the use of AI in human activities holds enormous advantages, challenges such as moral values, behaviors, trust, and privacy trail the deployment of the technology (Ahmad et al., 2023). Ethical issues and concerns in AI include innovation cost, consent issues, personal data misuse, criminal and malicious use, freedom and autonomy loss, and the loss of human decision-making, among others (Mengidis et al., 2019; Stahl, 2021a, 2021b). AI enhances organizational information security, fosters competitive advantage, and enhances customer relationships (Ahmad et al., 2023; Sayed & Muhammad, 2015; Rasheed et al., 2015). However, there are concerns that by 2030, the AI revolution will not only enhance societal progress and social control but will also raise ethical concerns, thus fueling growing divergent opinions on the positive impact on life and moral standing (Muller, 2020; Rainie et al., 2021).

2.4 Sustainability and Sustainable Development

In academic and scientific contexts, the concepts of *sustainable development* and *sustainability* are often intertwined, with the terms used interchangeably in much of the literature (Ruggerio, 2021). Although there is no single, universally accepted definition of *sustainable development*, the World Commission on Environment and Development (WCED, 1987) offered one of the most influential descriptions: *"development that meets the needs of the present without compromising the ability of future generations to meet their own needs."* In essence, sustainability represents the action-oriented dimension of sustainable development, emphasizing the translation of principles into practice (CEE, 2007). Sustainable development is a dynamic process that seeks to ensure the responsible use of resources, the strategic direction of investments, and the alignment of technological innovation and institutional transformation (Tremblay et al., 2020). Together, these elements must operate in harmony to enhance both present and future capacities to satisfy human needs and aspirations (Oti & Otalor, 2019; WCED, 1987).

While *sustainability* focuses on maintaining equilibrium between society's environmental use and the well-being of current and future generations, *sustainable development* specifically integrates economic growth with environmental stewardship (Oti & Otalor, 2019). It promotes a holistic approach that recognizes that prosperity and planetary health are interdependent (Tremblay et al., 2020). Sustainable development, therefore, entails achieving a careful balance between humanity's pursuit of improved living standards and well-being, and the preservation of resources upon which both present and future generations rely (IMF, 2002). As Lee et al. (2000) and Potter et al. (2004) highlight, it is about advancing human progress while ensuring that the natural resources and the ecosystems that sustain life remain intact. Sustainable development thus represents not just an environmental goal, but a moral and practical commitment to shared, long-term prosperity.

2.5 Dimensions of Sustainability and Sustainable Development

There are three dimensions of sustainability viz: the social, economic, and environmental perspectives, often referred to as the three pillars or triple bottom line. Sustainable development is achieved when these three dimensions—social, economic, and environmental—interact harmoniously. True sustainability requires balancing human welfare and economic growth with ecological preservation. This integrated approach is often represented through frameworks like the United Nations Sustainable Development Goals (SDGs), which connect global priorities across all three pillars.

2.5.1 Environmental Dimension

The environmental dimension focuses on maintaining the health and integrity of the planet's natural systems. It emphasizes responsible management of resources—such as water, soil, air, and biodiversity—to ensure that ecological processes are sustained for future generations. This involves reducing pollution, mitigating climate change, promoting renewable energy, and encouraging conservation. In sustainable development, this dimension ensures that economic and social progress does not come at the cost of environmental degradation.

Environmental factors are fundamental to achieving sustainability, compelling nations and businesses alike to allocate resources toward improving environmental performance and advancing SDGs related to ecological quality and public health. Lucas et al. (2019) argues that achieving the SDG target on reducing child mortality by 2030 is significantly impeded by persistently high death rates in Sub-Saharan Africa and South Asia—regions where environmental health risk factors such as poor air quality, unsafe water, and inadequate sanitation remain widespread. This underscores the urgent need to improve environmental conditions as a pathway to advancing health-related SDGs, particularly SDG 3 (Good Health and Well-being) and SDG 6 (Clean Water and Sanitation).

For **African countries,** enhancing environmental management offers multifaceted benefits. Investing in clean energy and sustainable waste management can reduce pollution and respiratory diseases, while improving access to clean water and sanitation directly lowers child and maternal mortality. Strengthening environmental regulations and promoting eco-innovation can also stimulate green industries, creating employment opportunities and fostering economic diversification. Moreover, protecting forests and biodiversity supports climate resilience, food security, and sustainable livelihoods—key to achieving SDG 13 (Climate Action) and SDG 15 (Life on Land). By integrating environmental sustainability into national development strategies, African nations can not only safeguard public health but also unlock long-term economic and social gains rooted in resilience and resource efficiency.

2.5.2 Economic Dimension

The economic dimension aims to foster long-term economic growth and prosperity without exhausting natural resources or increasing inequality. It focuses on creating systems that support innovation, employment, fair trade, and responsible production and consumption. Sustainable economic practices encourage efficiency, circular economy models, and investments that promote social and environmental well-being. In essence, this pillar seeks to balance profit with responsibility. Studies such as Islam et al.(2024) and Thore and Tarverdyan (2022), have shown that factors like health expenditure, household consumption, and unemployment significantly influence maternal and infant mortality rates, highlighting the essential role of economic investment in health for sustainable development. Therefore, increasing health expenditure and addressing macroeconomic determinants in Africa, are essential for achieving SDG health targets by 2063 (Adegoke et al., 2022; Islam, 2025).

2.5.3 Social Dimension

The social dimension centers on human well-being, equity, and inclusion. It involves ensuring access to basic needs such as education, healthcare, housing, and justice, while also promoting participation, cultural diversity, and community resilience. Sustainable development in this context seeks to create societies that are fair, safe, and inclusive, allowing both present and future generations to thrive.

Social considerations play a pivotal role in enabling business organizations to contribute effectively to a country's progress toward achieving the Sustainable Development Goals (SDGs) (Siakwah et al., 2019). Central to this is social governance, which encompasses the policies, practices, and ethical standards through which businesses engage with stakeholders to promote their well-being, protect their rights, and enhance shared prosperity (Islam, 2025). Strong social governance ensures that corporate activities are aligned with broader societal objectives, reinforcing trust and accountability. By cultivating meaningful relationships with stakeholders and safeguarding their interests, businesses not only strengthen their social license to operate but also make substantial contributions to the realization of multiple SDGs, particularly those related to decent work, reduced inequalities, and responsible production (Li et al., 2018).

Various social variables shape a nation's trajectory toward achieving the Sustainable Development Goals (SDGs), including life expectancy, population density, and workforce demographics. Longer life expectancies—often associated with improved healthcare systems—are pivotal in advancing health-related SDGs. Hao et al. (2020) emphasize that enhanced healthcare systems not only extend longevity but also reduce mortality rates and improve overall quality of life, contributing directly to SDG 3 (Good Health and Well-being). Increased access to healthcare services, vaccination programs, and preventive care can yield substantial improvements in public health outcomes, thereby accelerating progress toward health-related targets.

In the context of African countries, investing in robust and equitable healthcare systems can yield transformative benefits. Expanding community health infrastructure, strengthening maternal and child health services, and leveraging digital health innovations (such as telemedicine and mobile health platforms) can help bridge access gaps between urban and rural populations. These efforts not only enhance life expectancy but also build resilience against health crises, fostering a healthier and more productive workforce essential for sustainable economic growth.

Population density—particularly in rapidly urbanizing African cities—also plays a dual role in shaping sustainable development outcomes, aligning with SDG 11 (Sustainable Cities and Communities). High population density can intensify challenges such as overcrowding, pressure on infrastructure, and pollution. Yet, it also offers unique opportunities for resource efficiency, affordable housing development, and the advancement of public transportation systems. For African nations, sustainable urban planning presents a pathway to inclusive growth. By adopting strategies such as expanding green spaces, investing in clean energy solutions, and deploying smart city technologies, governments can transform urban centers into engines of innovation and sustainability.

Ultimately, addressing these social variables through integrated policies—linking health, urban planning, and demographic management—can help African countries not only accelerate SDG progress but also build resilient societies capable of sustaining inclusive and environmentally conscious growth.

2.6 AI in Africa's Sustainability Agenda

2.6.1 AGENDA 2063: The Africa We Want

Agenda 2063 is Africa's comprehensive blueprint and master plan for transforming the continent into a global powerhouse over 50 years (2013–2063). It serves as the strategic framework for achieving inclusive and sustainable development, reflecting Africa's collective aspirations for unity, self-determination, freedom, progress, and prosperity under the ideals of Pan-Africanism and the African Renaissance (Ndizera & Muzee, 2018; AUC, 2020; 2015). The origin of Agenda 2063 lies in the recognition by African leaders of the need to shift focus from past struggles—such as liberation and the fight against apartheid, which had defined the agenda of the Organisation of African Unity (OAU)—toward a new emphasis on social and economic transformation, continental integration, good governance, and peace and security. This reorientation aims to position Africa as a key global player capable of shaping its own destiny.

Fig. 4.4 Agenda 2063—"The Africa We Want": Technology and Innovation as Drivers of Inclusive Growth, Good Governance, Cultural Renaissance, and Peaceful and Secure Societies

During the 50th Anniversary of the OAU/AU in May 2013, African heads of state adopted the 50th Anniversary Solemn Declaration, reaffirming their commitment to a new era of development. This declaration enshrined the Pan-African Vision of "an integrated, prosperous, and peaceful Africa, driven by its own citizens and representing a dynamic force in the global arena." (African Union, 2013). Agenda 2063 is the roadmap for realizing this vision. Former AU Commission Chairperson Dr. Nkosazana Dlamini-Zuma captured this vision in a symbolic letter portraying the Africa of the future (AUC, 2015).

The long-term, 50-year horizon of Agenda 2063 is essential for addressing Africa's structural transformations and leveraging new opportunities. It recognizes progress in peace, stability, and economic growth; the importance of people-centered development, gender equality, and youth empowerment; and the influence of global changes, including technological advances and globalization. Furthermore, Africa's growing unity and potential in strategic sectors—including agriculture, infrastructure, education, health, and value addition to raw materials—provide strong foundations for sustainable growth (Thusi et al., 2024; Lopes et al., 2017; Bello-Schünemann et al., 2018). Agenda 2063 outlines Africa's Aspirations for the Future and establishes Flagship Programmes to accelerate economic growth and integration. It also includes 10-year Implementation Plans that define concrete actions, milestones, and outcomes to ensure that its goals are both quantitative and qualitative (UNECA, 2017). Ultimately, Agenda 2063 envisions a transformed, self-reliant, and globally influential Africa, empowered by its people and resources (AUC, 2015; Ndizera & Muzee, 2018).

2.6.2 AI's Role in the African Sustainability Agenda

Artificial intelligence (AI) can affect sustainable development. It can have a significant influence on the economic, social, and environmental perspectives of sustainability, and specific Sustainable Development Goals (SDGs) (Mubangizi, 2024). AI can boost productivity by driving innovation (such as new product development) and enhancing employment opportunities through the creation of new jobs.

2.6.2.1 AI Tools in the African Sustainability Agenda

2.6.2.1.1 Machine Learning (ML)

This core sub-area involves developing algorithms that allow computers to automatically learn from data and improve their performance over time without explicit programming. It includes subfields such as supervised learning, unsupervised learning, and reinforcement learning. To achieve digital transformation, Africa needs to leverage machine learning to enhance predictive analytics across health, agriculture, and governance (Idhalama & Makori, 2024). ML models can assist in yield prediction and pest control in agriculture, while in healthcare they support disease forecasting (Tapo et al., 2024). Financial inclusion can also be achieved through mobile credit scoring systems powered by ML for the unbanked (Ajuzieogu, 2017). However, the continent needs to quickly address the challenges of data scarcity and model bias (Khan et al., 2024).

2.6.2.1.2 Natural Language Processing (NLP)

NLP focuses on the interaction between computers and human language, enabling machines to understand, interpret, and generate human language. Applications range from language translation and sentiment analysis to virtual assistants and chatbots. NLP can support Africa's linguistic diversity by powering local-language translation, voice assistants, and sentiment analysis tools (Ajuzieogu, 2017). For example, the *Masakhane* project typifies African-language modeling efforts, which have considerably reduced reliance on Western-based datasets. As posited by Khan et al., (2024), it is expected that widespread adoption in the continent would strengthen education, public services, and cultural preservation.

2.6.2.1.3 Computer Vision

This field enables machines to "see" and interpret visual data from images and videos, similar to human vision. Key applications include facial recognition, medical image analysis, and enabling self-driving cars to identify obstacles. Africa could benefit from Computer Vision applications like medical imaging and precision agriculture. Image analysis aids early detection of diseases like malaria, while satellite-based monitoring optimizes irrigation (Arsalan et al., 2025). However, the need for ethical oversight in using biometrics for surveillance cannot be overemphasized (Tapo et al., 2024).

2.6.2.1.4 Robotics

This is an interdisciplinary branch of science that combines engineering and AI to design and build robots capable of performing tasks autonomously or semi-autonomously. AI-powered robots are used in manufacturing, medical surgery, and exploration. In Africa, robots can be deployed in mining, manufacturing, and healthcare. This can mitigate labor shortages and enhance workplace safety. AI-driven robots enable semi-autonomous agricultural systems and industrial process optimization (Tarhebai, 2024). Yet, high costs and limited infrastructure can restrict its adoption in Africa.

2.6.2.1.5 Expert Systems

These programs simulate the decision-making ability of human experts by using a knowledge base and a set of rules (often "if-then" logic) to solve complex problems in specific domains like medical diagnosis or financial forecasting. Expert systems integrated with fuzzy logic enable adaptive diagnostic systems for uncertain environments with poor data, like most African countries (Peralta et al., 2025). The system can simulate human reasoning, which is valuable in rural healthcare and financial decision-making (Khan et al., 2024).

2.6.2.1.6 Deep Learning (DL)/Neural Networks

Deep learning is a specialized subset of machine learning that uses multi-layered artificial neural networks inspired by the human brain. It has driven significant progress in areas like image and speech recognition due to its ability to process vast amounts of data and identify complex patterns. Deep Learning aids urban mapping, energy forecasting, and autonomous systems, and it can drive breakthroughs in voice and image recognition across the African continent (Khan et al., 2024). For example, African research communities like *Deep Learning Indaba* (a grassroots movement bringing together African AI researchers and practitioners to build the future of AI in Africa) have been helping to accelerate regional expertise (Idhalama & Makori, 2024).

2.6.2.1.7 Fuzzy Logic

This approach deals with reasoning under uncertainty and imprecision, allowing AI systems to make decisions in scenarios where data is not strictly true or false. It is used in control systems for appliances like washing machines and air conditioners. Fuzzy logic can help Africans overcome critical resource constraints such as water management and renewable energy control (Allahviranloo et al., 2024). Its integration with expert systems can enhance adaptive control and energy optimization (Peralta et al., 2025).

2.6.2.1.8 Generative AI

An emerging and rapidly growing field, generative AI focuses on creating new content, such as text, images, or music, by learning from existing data patterns. Large language models (LLMs) like ChatGPT is an example. For Africa, generative AI presents both creative and economic opportunities, as it empowers local content creation, translation, and education, although it creates ethical issues such as misinformation and copyright (Ajuzieogu, 2017; Eyal, 2021). That is why Africa must continue to push for culturally-aligned datasets and responsible AI policies.

2.6.2.1.9 Big Data Analytics

Big data analytics (BDA) is defined as the process of using advanced analysis algorithms running on powerful supporting platforms to uncover potentials concealed in big data, such as hidden patterns or unknown correlations, by mining massive dataset at different levels in real-time or near real-time, including modeling, visualization, prediction, and optimization (Biesialska et al., 2020). It involves advanced technologies for analyzing large-scale heterogeneous datasets, big data mining, and statistical analysis, where datasets are broken down and executed across multiple nodes in parallel (Pramanik et al., 2020).

3 The Role of AI in Specific Sectors of African Economies

3.1 Agriculture and Food Security

Agriculture remains a major employer and source of livelihoods across Africa. AI-enabled tools have been promoted as a means to improve productivity, enhance climate resilience and strengthen food security. Applications include crop disease detection using computer vision, yield prediction and weather forecasting through machine learning, and decision-support tools for fertilizer application, irrigation and pest management. These systems can help smallholders anticipate climate shocks, reduce input waste and access advisory services, potentially contributing to SDGs related to poverty, hunger and climate action (Kuzmina et al., 2024; Mienye et al., 2024).

At the same time, AI in agriculture often depends on satellite imagery, mobile data and digital platforms controlled by foreign firms. This raises concerns about data ownership, the terms under which farmers' data is collected and monetized, and whether AI-driven value chains will benefit smallholders or primarily agribusiness and platform companies. Ethical questions thus center on data governance, informed consent, fair value distribution and the risk of deepening existing rural inequalities.

3.2 Health Systems and Public Health

AI is increasingly applied in African health systems for diagnostics, disease surveillance, resource allocation and health information management. Machine learning models have been developed for detecting tuberculosis, cervical cancer and malaria using medical imaging or rapid tests; AI-based chatbots and triage systems support telemedicine and digital health platforms; and predictive analytics are used for epidemic forecasting and health workforce planning (Mienye et al., 2024).

Such applications can augment scarce health resources, improve early detection and extend services to remote communities. Yet they also raise ethical concerns around data privacy, consent, algorithmic bias and the potential deskilling of health workers. If diagnostic models are trained on non-African or biased datasets, they may misclassify conditions in African populations, leading to unequal treatment outcomes. Moreover, the increasing reliance on private digital health platforms may fragment health systems and complicate public accountability.

3.3 Environmental Management and Climate Action

AI tools are deployed in environmental monitoring, climate modeling and disaster risk reduction. Remote sensing and machine learning combinations are used to track deforestation, land use change and water stress; early warning systems integrate AI to predict floods, droughts and cyclones; and optimization algorithms inform energy grid management and integration of renewables. These applications support climate adaptation and mitigation, a crucial priority for African countries that are highly vulnerable to climate impacts (Kuzmina et al., 2024).

However, the environmental footprint of AI infrastructure—data centers, communication networks and end-user devices—must also be considered. Concentration of data centers in regions with carbon-intensive electricity, or inadequate e-waste management, can undermine environmental goals and shift ecological burdens onto local communities.

3.4 Smart Cities, Governance, and Public Services

Urbanization is rapidly reshaping African societies, and AI is increasingly integrated into "smart city" and e-governance initiatives. AI-driven systems support traffic management, public transport optimization, digital identity and authentication, and in some cases predictive policing and video surveillance. Digital ID systems combined with AI can facilitate access to social services and financial inclusion, but also raise concerns about surveillance, exclusion of those lacking documentation, and risks of discrimination in law enforcement.

AI in governance holds potential to improve efficiency, transparency and responsiveness, yet may also entrench opaque decision-making if algorithms are proprietary or not explainable. In contexts where democratic oversight is fragile, the deployment of AI in security and surveillance can exacerbate human rights risks, particularly for activists, minorities and migrants (Mubangizi, 2024).

3.5 Economic Development and Financial Inclusion

African fintech firms have been early adopters of AI, particularly for credit scoring, fraud detection and customer service. Alternative data-based credit scoring models draw on mobile money transactions, airtime purchases, social networks and geolocation to assess creditworthiness, enabling access to loans for individuals and small enterprises previously excluded from formal banking. This aligns with SDGs on economic growth, decent work and reduced inequalities.

Nonetheless, opaque credit models risk embedding bias, punishing individuals for factors beyond their control, and locking them into cycles of debt. In addition, high interest rates on digital loans and aggressive debt collection practices have raised questions about consumer protection and financial ethics. These examples illustrate how AI can simultaneously advance and threaten sustainability objectives, depending on governance choices.

4 Ethical Risks and Challenges

Artificial Intelligence (AI) presents both transformative opportunities and profound ethical challenges for African nations. Its integration across governance, finance, and development sectors exposes long-standing structural inequalities and raises concerns over data sovereignty, algorithmic bias, and digital dependency.

Several structural features of African societies and economies heighten the importance of AI ethics. Regulatory and institutional capacity for digital technologies remains limited in many countries, with data protection authorities under-resourced and AI-specific regulations nascent or absent. Recent scholarship identifies these dynamics as part of a broader system of digital colonialism, where global technology infrastructures reproduce colonial hierarchies of power and knowledge (Salami, 2024; Regilme, 2024).

4.1 Data and Privacy

The digital transformation in Africa has occurred within largely underdeveloped data protection regimes. Therefore, weak data protection regimes remain a critical barrier to ethical AI governance in Africa. Although several countries—such as South Africa and Nigeria—have enacted data protection laws, enforcement remains inconsistent due to limited institutional capacity (Mubangizi, 2024). This regulatory gap facilitates mass data extraction and processing without informed consent, allowing corporations and governments to exploit user data for surveillance and commercial gain (Effoduh, 2025). As Bhanye (2025) notes, the continent has become a site of "cognitive extraction", where citizens serve as data providers rather than data owners.

The widespread deployment of biometric identification systems, smart city projects, and predictive policing technologies compounds privacy risks. Without transparent oversight mechanisms, these tools enable pervasive surveillance that can suppress dissent or target marginalized communities (Omotubora & Basu, 2024). The resulting erosion of privacy undermines not only individual rights but also the legitimacy of digital governance structures.

4.2 Bias, Discrimination, and Exclusion

AI models trained on datasets from the Global North often fail to represent African linguistic, cultural, and socioeconomic realities. This leads to systematic errors and exclusions, particularly in facial recognition, language processing, and credit scoring systems (Ayana et al., 2024; Arora et al., 2023). Algorithmic decisions based on such skewed data can reproduce existing patterns of inequality—along lines of gender, ethnicity, or geography—thereby entrenching social hierarchies rather than mitigating them (Hassan, 2023).

The exclusionary nature of digital systems also affects individuals without formal digital identities or online presence, who are often denied access to services tied to AI systems. Barrett et al. (2025) argue that data invisibility has become a new form of marginalization, deepening the digital divide. Addressing these biases requires a deliberate effort to embed African data ethics and representation into global AI development frameworks.

4.3 Power, Dependency, and "Digital Colonialism"

African countries face a new wave of colonization—technological dependency resulting from the dominance of the digital infrastructure by non-Africans. Africa's cloud computing, data storage, and communication platforms, are controlled by companies like Meta, Google, and Amazon leading to asymmetric power relations and reduced national autonomy (Effoduh, 2025; Ayana et al., 2024; Salami, 2024). Regilme (2024) describes this as a continuation of extractive colonial practices, where data rather than minerals become the resource of exploitation.

Holden and Harsh (2024) highlight how these dependencies are sustained through "infrastructural pipelines" that connect Africa's data labor to global AI industries. These dynamics not only limit local innovation but also centralizes control of African data in foreign jurisdictions, constraining the continent's ability to shape its digital future.

4.4 Labor and Socio-Economic Impacts

The adoption of AI across Africa is reshaping labor markets, exacerbating labor exploitation. The automation-induced proliferation of new forms of low-paid digital labor, data labeling, content moderation, and annotation continues to threatens traditional employment in logistics and manufacturing (Regilme, 2024). Changes in job structures, employee demands, working hours, pay, and social security. Furthermore, the data systems underlying AI connect historical patterns of colonialism, surveillance, and exploitation, showing how old power relations still shape present and future inequalities (Holden & Harsh, 2024). The economic benefits of AI thus accrue primarily to elites and multinational corporations, reinforcing socioeconomic disparities (Arora et al., 2023). Without robust labor protections and redistributive policies, the promise of AI-driven development risks becoming another mechanism of inequality and dependency.

4.5 Environmental and Infrastructural Concerns

AI's infrastructural demands pose significant environmental challenges. Data centers and machine learning operations require immense energy, straining Africa's limited and often carbon-intensive power grids (Regilme, 2024). Simultaneously, the proliferation of electronic devices contributes to rising e-waste, which is frequently dumped or informally recycled under hazardous conditions can damage the environment (Mariniello, 2025).

Scholars such as Bhanye (2025) and Kaur et al., (2025) posit that these issues highlight historical patterns of environmental exploitation in the Global South. Therefore, efforts to advance AI must be anchored on local environmental realities and synchronized with sustainability goals.

4.6 Governance and Institutional Weaknesses

AI governance in most African states continue to encounter challenges like weak institutional capacity, a lack of technical skills and regulatory frameworks needed to assess AI systems' ethical and societal implications (Omotubora & Basu, 2024). Inadequate legal frameworks create opaque algorithms and facilitate data exploitation. Corruption and lack of transparency in both public and private sectors further erode accountability. Ayana et al., (2024) call for a decolonized AI governance framework rooted in African culture to enable the continent to create its own standards of fairness, transparency, and social benefit, rather than importing Western-centric models that do not reflect local realities.

AI in Africa embodies both the promise of technological progress and the perils of renewed colonization. The issues of data exploitation, bias, labor precarity, and environmental degradation reveal how global AI infrastructures perpetuate structural inequalities. An independent and justice-oriented home-grown African data sovereignty, inclusive governance, and ethical accountability is essential for ensuring that AI contributes to equitable development rather than reproducing historical dominance of superpowers.

5 Policy and Governance Considerations

Achieving digital sovereignty in Africa necessitates a fundamental re-envisioning of data governance architectures that advance an inclusive, equitable, and ethically grounded AI economy (Akpudo et al., 2024; Ndemo & Thegeya, 2022). African policymakers must articulate and institutionalize data governance frameworks that are firmly anchored in African epistemologies, ethical traditions, and socio-cultural values, while simultaneously embedding human-centered evaluative criteria for the design, deployment, and oversight of AI systems across the continent (Akpudo et al., 2024). Such frameworks are essential to ensure that AI development and application are consistent with international and regional human rights standards, particularly those upholding privacy, equality, nondiscrimination, inclusivity, diversity, safety, fairness, transparency, accountability, and sustainable economic advancement, with special attention to the protection and empowerment of marginalized and vulnerable populations (United Nations High Commissioner for Human Rights [UNHCHR], 2021). Realizing these imperatives requires multi-scalar governance mechanisms, encompassing national policy reforms, regional harmonization, and continental cooperation to establish a coherent and contextually relevant AI ecosystem in Africa.

5.1 National Level

The need for strong data protection measures has never been more pressing in an age where personal data has replaced natural resources as a primary driver of economic value (KPMG, 2023). As Adewumi (2022) observes, transborder flow of personal data has evolved into a key aspect of global trade and the lifeblood of digital capitalism, positioning data governance at the heart of today's geopolitical and economic negotiations . Therefore, African states must recalibrate data governance by enacting and reinforcing comprehensive data protection laws and developing AI-specific legislations and digital governance strategies. These should be supported by the establishment of independent ethics oversight institutions and clear standards on transparency and impact assessment. Through participatory co-governance and multi-stakeholder collaboration, Africa can construct a normatively grounded and resilient data governance ecosystem, one capable of mediating the tensions between innovation, rights protection, and digital sovereignty in an increasingly data-driven global order.

5.2 Regional and Continental Level

At the regional and continental levels, the ECOWAS, SADC, EAC and the African Union can institutionalize pan-African regulatory framework that align the continent's sociocultural realities with global data ethics. Initiatives such as the AU Digital Transformation Strategy (2020–2030) and the emerging Continental Artificial Intelligence Strategy can drive the process to ensure cooperation cross-border data governance, shared regulatory sandboxes and capacity building, and joint bargaining with global tech companies. Such strategies can mitigate the asymmetrical power dynamics that often characterize Africa's engagement with global data economies and to reduce reliance on Euro-American epistemologies that dominate AI discourse. These actions are necessary to enable Africa reclaim digital agency and ensure that the African AI landscape evolves as a site of ethical pluralism, inclusive innovation, and socioeconomic emancipation.

5.3 Sectoral/Technical Measures

A multi-scalar data governance framework for Africa must encompass a comprehensive ethical architecture that integrates context-sensitive design principles and systematic impact assessments throughout the lifecycle of AI systems. Such a framework should establish sector-specific guidelines for high-risk domains—particularly in healthcare, policing, and finance—where algorithmic decisions have profound social, economic, and ethical implications. Central to this model is the institutionalization of algorithmic transparency, auditability, and accountability, ensuring that AI systems remain subject to continuous scrutiny and public oversight. Equally critical is the development of accessible redress and remediation mechanisms for individuals or groups adversely affected by AI deployment, thereby operationalizing the principles of justice and fairness in digital governance. Collectively, these measures will advance the responsible, equitable, and human-cantered development of AI within the African context, fostering a governance paradigm that balances innovation with ethical integrity and technological progress with social protection.

5.4 Capacity Building and Inclusion

Africa's aspiration to develop a data governance framework grounded in indigenous epistemologies, ethical traditions, and sociocultural values will remain elusive unless it strategically reconfigures its technological development agenda to prioritize local capacity building and epistemic autonomy. This necessitates systematic investment in universities, research institutes, and AI-driven startups, as well as the expansion of digital and computational infrastructure capable of supporting advanced innovation ecosystems. Building endogenous AI expertise is not merely a technical imperative but a political and epistemological act—a necessary step toward reclaiming agency over how data and technology shape African futures. Equally, ensuring gender and youth inclusion within AI policymaking and discourse must occupy a central position in this transformative agenda, recognizing that sustainable AI ecosystems depend on diverse participation, representation, and leadership. Furthermore, the development of AI systems such as natural language processing (NLP) must be deliberately oriented toward linguistic plurality, ensuring that technologies are designed to preserve and empower local languages rather than marginalize them. Through these interventions, Africa can cultivate a decolonial AI landscape that not only aligns with global ethical standards but also reflects and amplifies its own cultural and intellectual heritage.

6 Case Examples

The African Cassava Agronomy Initiative (ACAI), coordinated by the International Institute of Tropical Agriculture (IITA), exemplifies the growing integration of AI-driven technologies into Africa's agricultural transformation agenda. Its mobile agronomic advisory platform, Akilimo, provides cassava growers with context-specific recommendations on fertilizer application, weed management, optimal planting techniques, and intercropping strategies—such as cassava with maize in Nigeria and sweet potato in Tanzania (Abioye et al., 2024). By September 2022, Akilimo had been deployed across 370,000 hectares in Nigeria and Tanzania, reaching over 809,000 farmers and training 6,504 extension agents, with more than 75% of users reporting improvements in yield and profit (Akilimo, 2022; Ampadu-Boakye et al., 2025; Laborde, 2022). While these outcomes underscore the transformative potential of AI in agricultural productivity, they also expose ethical and governance challenges related to data ownership, algorithmic transparency, and digital dependency. Farmers' reliance on proprietary agronomic algorithms raises questions about who controls agricultural data and how algorithmic recommendations align with local knowledge systems and environmental sustainability goals.

Similar tensions manifest in AI applications in health and fintech sectors. In Nigeria's health system, for instance, AI-powered diagnostic tools such as Ubenwa, which uses machine learning to detect birth asphyxia through infant cry analysis, demonstrate significant potential for early detection and life-saving interventions. However, such innovations also surface ethical risks concerning informed consent, data privacy, and bias in training datasets, which may reinforce systemic inequities in access to care. In fintech, platforms like Carbon and Kuda Bank employ AI for credit scoring and fraud detection, expanding financial inclusion for previously unbanked populations. Yet, these same systems may inadvertently perpetuate exclusion through biased credit algorithms or opaque decision-making processes, raising governance concerns over algorithmic accountability and consumer protection.

Collectively, these cases illustrate that while AI holds immense potential to accelerate socioeconomic development in Nigeria and across Africa, its deployment must be embedded within robust ethical and regulatory frameworks. Such frameworks should mandate algorithmic transparency, participatory oversight, localized data stewardship, and clear redress mechanisms for those adversely affected. Without these safeguards, the continent risks replicating digital dependency and data colonialism, rather than realizing a decolonial, human-cantered AI governance model tailored to Africa's epistemic and sociocultural realities.

7 Conclusion and Recommendations

As Africa pursues the attainment of the sustainable development goals through its Agenda 2063, AI presents a powerful opportunity for agricultural revolution, revamping health and other infrastructure, developing a robust resistance to climate change and upgrade governance systems. However, the continent must recognize the risks associated with AI development and adoption such as data exploitation, labor precarity, algorithmic discrimination, and ecological damages, including structural inequalities, technological dependencies and environmental vulnerabilities. Therefore, Africa must prioritize building local technologies anchored on intellectual freedom, thereby achieving self-governance in AI development, deployment and use. This will require sustained focus on equity, justice and respect for human rights. Therefore, this chapter makes the following recommendations:

For Africa to meaningfully benefit from artificial intelligence, the continent must move beyond its current position as a passive supplier of raw data to become an active co-producer of AI technologies. Achieving this transition requires sustained investment in indigenous AI research and development, human capital formation through education and skills development, and the expansion of digital and computational infrastructure. Strengthening local innovation ecosystems will not only reduce dependence on foreign technologies but also enhance Africa's capacity to shape AI systems that respond to its developmental priorities and sociocultural realities.

The local design, deployment, and use of AI systems in Africa—including data generation, collection, and processing—should be firmly grounded in principles of justice, respect for human rights, and environmental sustainability. Embedding ethical considerations at every stage of the AI lifecycle is essential to preventing harm, mitigating bias, and ensuring that technological advancement aligns with broader societal values and sustainable development goals. This approach underscores the need for responsible AI frameworks that are context-sensitive and socially accountable.

AI deployment in Africa must adopt a holistic and inclusive research agenda that systematically considers its social, economic, and environmental implications. Such an agenda requires robust interdisciplinary collaboration among computer scientists, social scientists, legal scholars, ethicists, and environmental researchers. By integrating diverse perspectives and methodologies, African AI research can generate more comprehensive evidence to inform policy, anticipate unintended consequences, and support equitable and sustainable technological adoption.

The development of inclusive AI governance in Africa depends on the establishment of strong national regulatory frameworks, complemented by coordinated regional and continental governance mechanisms. These institutions must be capable of overseeing AI development while ensuring transparency, accountability, and public trust. Crucially, inclusive governance requires the meaningful participation of civil society organizations, local communities, labor groups, and vulnerable populations in decision-making processes related to AI deployment. Such participatory governance structures are essential for aligning AI systems with public interest and safeguarding social inclusion.

References

Abioye, D. O., Popoola, O., Akande, A., Fadare, D. A., Omitoyin, S. A., Yinusa, B., & Kolade, O. O. (2024). Farmers' willingness to adopt digital application tools in Ogun State, Nigeria. *Journal of Strategy and Management*, Vol. ahead-of-print No. ahead-of-print. https://doi.org/10.1108/JSMA-06-2023-0135

Access Now. (2018). *Human rights in the age of artificial intelligence*. Access Now.

Adegoke, Y. O., Mbonigaba, J., & George, G. (2022). Macro-economic determinants, maternal and infant SDG targets in Nigeria: Correlation and predictive modeling. *Frontiers in Public Health, 10*, 999514. https://doi.org/10.3389/fpubh.2022.999514

Adewumi, A. (2022). *Adequate protection: An analysis of Nigeria's data protection laws within an emerging global data protection framework* [Unpublished master's thesis]. University of Victoria.

African Union. (2013). *50th Anniversary Solemn Declaration*. AU.

African Union Commission (AUC). (2015). *Agenda 2063: The Africa We Want*. AUC.

African Union Commission (AUC). (2020). *Agenda 2063 and SDG implementation*. African Peer Review Mechanism.

Ahmad, S. F., Han, H., Alam, M. M. Rehmat, M. K., Irshad, M., Arraño-Muñoz, M., & Antonio Ariza-Montes, A. (2023). Impact of artificial intelligence on human loss in decision making, laziness and safety in education. *Humanities and Social Sciences Communication, 10*(311), 1–14. https://doi.org/10.1057/s41599-023-01787-8

AI Ethics Impact (AIEI) Group. (2020). *From principles to practice—An interdisciplinary framework to operationalise AI ethics*. VDE/Bertelsmann Stiftung.

AI High-Level Expert Group (AI HLEG). (2020). *Sectorial considerations for trustworthy AI—Taking AI's context specificity into account*. European Commission.

Ajuzieogu, U. (2017). Cross-disciplinary AI impact assessment for African contexts: A comprehensive research framework. Retrieved December 16, 2025, from https://doi.org/10.13140/RG.2.2.26641.34406

Akilimo (2022), available at: https://akilimo.org/ (accessed December 2025).
Akpudo, U. E., Effoduh, J. O., Kong, J. D., & Yongsheng Gao, Y. (2024). Unveiling AI concerns for Sub-Saharan Africa and its vulnerable groups. *International Conference on Intelligent and Innovative Computing Applications (ICONIC), 7th-9th November 2024, Pearle Beach Resort & Spa, Flic en Flac, Mauritius,* 45–55. https://doi.org/10.59200/ICONIC.2024.007
Allahviranloo, T., Pedrycz, W., & Seyyedabbasi, A. (2024). *Decision-making models: A perspective of fuzzy logic and machine learning*. Academic Press.
Ampadu-Boakye, T., Ogunsanmi, T., Dotse, S. K., Fasakin, J., Mwantumu, O., Mensah, N. A., Kreye, C., & Baijukya, F. (2025). *AKILIMO Progress Report*. CGIAR Sustainable Farming.
Arora, A., Barrett, M., Lee, E., Oborn, E., & Prince, K. (2023). Risk and the future of AI: Algorithmic bias, data colonialism, and marginalization. *Information and Organization, 33*(3), 100478. https://doi.org/10.1016/j.infoandorg.2023.100478
Arsalan, M. H., Mubin, O., Al Mahmud, A., Khan, I. A., & Hassan, A. J. (2025). Mapping data-driven research impact science: The role of machine learning and artificial intelligence. *Metrics, 2*(2), 5. https://doi.org/10.3390/metrics2020005
Ayana, G., Dese, K., Nemomssa H. D., Habtamu, B., Mellado, B., Badu, K., Yamba, E., Faye, S. L., Ondua, M., Nsagha, D., Nkweteyim, D., & Kong, J. D. (2024). Decolonizing global AI governance: assessment of the state of decolonized AI governance in Sub-Saharan Africa. *Royal Society Open Science, 11*(8), 231994.
Babuta, A., Oswald, M., & Janjeva, A. (2020). *Artificial intelligence and UK National Security—Policy considerations*. Royal United Services Institute for Defence and Security Studies.
Barrett, T., Okolo, C. T., Biira, B., Sherif, E., & Zhang, A. (2025). African data ethics: A discursive framework for Black decolonial AI. *Proceedings of the 2025 ACM Conference on Fairness, Accountability, and Transparency* (June 23–26, Athens, Greece), 334–349. https://doi.org/10.1145/3715275.3732023
Bello-Schünemann, J., Cilliers, J., Donnenfeld, Z., Aucoin, C., & Porter, A. (2018). African Futures: Key trends to 2035. *Journal of Futures Studies, 23*(1), 127–140.
Bengio, Y., Lecun, Y., & Hinton, G. (2021). Deep Learning for AI. *Communications of the ACM, 64*(7), 58–65. https://doi.org/10.1145/3448250
Bentham, J. (1789). *An introduction to the principles of morals and legislation*. Dover Publications Inc.
Berger, K. M., & Roderick, J. (2014). *National and transnational security implications of big data in the life sciences*. American Association for the Advancement of Science.
Bhanye, J. (2025). *Artificial Intelligence (AI) and the new colonialism of climate data in the Global South: Toward just and decolonial adaptation futures*. Public Affairs Research Institute Working Paper. https://pari.org.za/wp-content/uploads/2025/12/20251205_AI_Johannes-Bhanye_WorkingPaper.pdf
Biesialska, K., Franch X., & Muntes-Mulero V. (2020). Big data analytics in Agile software development: A systematic mapping study. *Information and Software Technology, 132*, 106448.
Birhane, A. (2020). Algorithmic colonization of Africa. In S. Cave, & K. Dihal (Eds.), *Imagining AI: How the world sees intelligent machines* (pp. 247–260). Oxford University Press.
Boden, M. A. (2018). *Artificial Intelligence: A very short introduction*. OUP Oxford.
Bowie, N. E. (1999). Business ethics: *A Kantian perspective*. Blackwell Publishers.
Centre for Data Ethics and Innovation. (2019). *Interim report: Review into Bias in algorithmic decision-making*. Centre for Data Ethics and Innovation.
Centre for Environment Education. (2007). *Sustainable development* (Vol. 1). Internship Series.
Childress, J. F., & Beauchamp, T. L. (1979). *Principles of biomedical ethics*. Oxford University Press.
Coeckelbergh, M. (2020). *AI ethics*. The MIT Press.
Coleman, D. (2019). Digital Colonialism: The 21st century scramble for Africa through the extraction and control of user data and the limitations of Data Protection Laws. *Michigan Journal of Race & Law, 24*(2), 417–439. https://doi.org/10.36643/mjrl.24.2.digital
da Costa, A., & Freita, Z. C. (2025). Digital ethics and the future of education: An urgent dialogue. *Journal of Educational Technology Innovation and Applications, 1*(1), 27–34.
De George, R. T. (1999). *Business ethics* (5th ed.). Prentice Hall.

de Reuver, M., van Wynsberghe, A., Janssen, M., & van de Poel, I. (2020). Digital platforms and responsible innovation: Expanding value sensitive design to overcome ontological uncertainty. *Ethics and Information Technology*, *22*, 257–267 https://doi.org/10.1007/s10676-020-09537-z

Dignum, V. (2019). *Responsible Artificial Intelligence: How to develop and use AI in a responsible way* (1st ed.). Springer.

Effoduh, J. O. (2025). Digital colonialism and the role of local intermediaries: examining big tech's impact on data sovereignty and human rights in Africa. *Business and Human Rights Journal*, *10*(2–3), 301–317. https://doi.org/10.1017/bhj.2025.21

Eke, D. O. (2023). ChatGPT and the rise of generative AI: Threat to academic integrity?. *Journal of Responsible Technology*, *13*, 100060. https://doi.org/10.1016/j.jrt.2023.100060

Eke, D. O., Wakunuma, K., & Akintoye, S. (Eds.). (2023). *Responsible AI in Africa, social and cultural studies of robots and AI.* Palgrave Macmillan. https://doi.org/10.1007/978-3-031-08215-3_1

Eke, D., Stahl, B. (2024). Ethics in the governance of data and digital technology: An analysis of European data regulations and policies. *Digital Society*, *3*(11), 1–23. https://doi.org/10.1007/s44206-024-00101-6

European Data Protection Supervisor (EDPS). (2020). *EDPS opinion on the European Commission's White Paper on artificial intelligence—A European approach to excellence and trust (Opinion 4/2020).* EDPS.

Eyal, M. (2021). Mapping the AI Landscape. University of Alabama Legal Studies Research Paper No. 4033597, Available at SSRN: https://ssrn.com/abstract=4033597

Floridi, L., & Taddeo, M. (2016). What is data ethics?. *Philosophical Transactions of the Philosophical Transactions: Mathematical, Physical and Engineering Sciences*, *374*(2083).

Floridi, L., Cowls, J., Beltrametti, M., Chatila, R., Chazerand, P., Dignum, V., Luetge, C., Madelin, R., Pagallo, U., Rossi, F., Schafer, B., Valcke, P., & Vayena, E. (2018). AI4 People—An ethical framework for a good AI society: Opportunities, risks, principles, and recommendations. *Minds & Machines 28*, 689–707. https://doi.org/10.1007/s11023-018-9482-5

Fothergill, B. T., Knight, W., Stahl, B. C., & Ulnicane, I. (2019). Responsible data governance of neuroscience big data. *Frontiers in Neuroinformatics*, *13*(28), 426225. https://doi.org/10.3389/fninf.2019.00028

Guenduez, A. A., Walker, N., & Demircioglu, M. A. (2025). Digital ethics: Global trends and divergent paths. *Government Information Quarterly*, *42*(3), 1–20. https://doi.org/10.1016/j.giq.2025.102050

Gupta, A., & Heath, V., 2020. AI ethics groups are repeating one of society's classic mistakes. *MIT Technology Review*. Retrieved December 9, 2025 from https://www.technologyreview.com/2020/09/14/1008323/ai-ethics-representation-artificial-intelligence-opinion/

Haenlein, M., & Kaplan, A. (2019). A brief history of artificial intelligence: On the past, present, and future of artificial intelligence. *California Management Review*, *61*, 5–14.

Hagerty, A., & Rubinov, I. (2019). Global AI ethics: A review of the social impacts and ethical implications of artificial intelligence. *arXiv*:1907.07892. https://arxiv.org/abs/1907.07892

Hall, W., & Pesenti, J. (2017). *Growing the artificial intelligence industry in the UK*. Department for Digital, Culture, Media & Sport and Department for Business, Energy & Industrial Strategy.

Hao, L., Xu, X., Dupre, M. E., Guo, A., Zhang, X., Qiu, L., . . . Gu, D. (2020). Adequate access to healthcare and added life expectancy among older adults in China. *BMC Geriatrics*, *20*(1). https://doi.org/10.1186/s12877-020-01524-9

Hassan, Y. (2023). Governing algorithms from the South: A case study of AI development in Africa. *AI & Society*, *38*, 1429–1442. https://doi.org/10.1007/s00146-022-01527-7

Holden, K., & Harsh, M. (2024). On pipelines, readiness and annotative labour: Political geographies of AI and data infrastructures in Africa. *Political Geography*, *113*, 103150. https://doi.org/10.1016/j.polgeo.2024.103150

Idhalama, O. U., & Makori, E. O. (2024). Artificial intelligence, deep learning, machine learning, robotics and digital transformation: applications, implications and future. *Ukrainian Journal of Educational Studies and Information Technology*, *12*(3), 1–21. https://doi.org/10.32919/uesit.2024.03.01

International Monetary Fund (IMF) (2002). *Assessing Sustainability: Methodological Issues*, (28 May). Washington DC: International Monetary Fund.

Islam, H. (2025). Nexus of economic, social, and environmental factors on sustainable development goals: The moderating role of technological advancement and green innovation. *Innovation and Green Development*, *4*(1), 100183. https://doi.org/10.1016/j.igd.2024.100183

Islam, H., Soumia, L., Rana, M., Madavarapu, J. B., & Saha, S. (2024). Nexus between perception, purpose of use, technical challenges and satisfaction for mobile financial services: Theory and empirical evidence from Bangladesh. *Technological Sustainability*, *3*(2), 147–170. https://doi.org/10.1108/techs-10-2023-0040

Javaid, M., Haleem, A., Khan, I. H., & Suman, R. (2023). Understanding the potential applications of artificial intelligence in agriculture sector. *Advanced Agrochem*, *2*(1), 15–30.

Kant, I. (1789). *Critique of practical reason*. Reclam.

Kant, I. (1797). *Groundwork of the metaphysics of morals*. Reclam.

Kaur, A., & Verma, S. S. (2025). Understanding India's Role in International Climate Negotiations from a Social Constructivist Perspective. *India Review, 24*(1), 36–54.

Khan, I. U., Ouaissa, M., Fayaz, M., & Ullah, R. (Eds.). (2024). *Artificial intelligence for intelligent systems: Fundamentals, challenges, and applications*. Taylor & Francis Group. https://doi.org/10.1201/9781003496410

KPMG. (2023). *Nigeria Data Protection Act 2023 Review*. KPMG Advisory Services.

Kuzmina, T., Podbiralina, G., & Baburina, O. (2024). Impact of AI on sustainable development in Africa. *E3S Web of Conferences*, *583*, 08005. https://doi.org/10.1051/e3sconf/202458308005

Laborde, D. (2022). *Critical food and fertilizer price increase and its impact on smallholder farmers in Africa*. TICAD 8.

Latorre, L., Rego, E., & de Leo, L. (2025). *Tech report: Digital ethics*. Inter-American Development Bank.

Lee, K., Holland, A., & McNeill, D. (Eds.). (2000). *Global sustainable development in the twenty-first century*. Edinburgh University Press.

Li, L., Xia, X., Chen, B., & Sun, L. (2018). Public participation in achieving sustainable development goals in China: Evidence from air pollution control practice. *Journal of Cleaner Production*, *201*, 499–506. https://doi.org/10.1016/j.jclepro.2018.08.046

Lopes, C., Hamdok, A., & Elhiraika, A. (2017). An overview of structural transformation and macroeconomic policy imperatives in Africa. In C. Lopes, A. Hamdok, & A. Elhiraika (Eds.), *Macroeconomic policy framework for Africa's structural transformation* (pp. 1–35). Palgrave Macmillan.

Lucas, P. L., Hilderink, H. B. M., Janssen, P. H. M., Samir K. C., Vuuren, D. P., & Louis Niessen, L. (2019). Future impacts of environmental factors on achieving the SDG target on child mortality—A synergistic assessment. *Global Environmental Change*, *57*, 101925. https://doi.org/10.1016/j.gloenvch a.2019.05.009

MacIntyre, A. C. (2007). *After virtue: A study in moral theory*. University of Notre Dame Press.

Mariniello, M. (2025). Colonialism 4.0 and Coloniality 4.0. *AI & Society*, 1–15. https://doi.org/10.1007/s00146-025-02736-6

Mayor, A. (2018). *Gods and robots: Myths, machines, and ancient dreams of technology*. Princeton University Press.

McCarthy, J. (2007). What is artificial intelligence?. Retrieved December 10, 2025, from https://www-formal.stanford.edu/jmc/whatisai.pdf

Mengidis, N., Tsikrika, T., Vrochidis, S., & Kompatsiaris, I. (2019). Blockchain and AI for the next generation energy grids: Cybersecurity challenges and opportunities. *Information & Security*, *43*(1), 21–33. https://doi.org/10.11610/isij.4302

Metcalf, J., Keller, E. F., & Boyd, D. (2016). *Perspectives on big data, ethics, and society*. Council for Big Data, Ethics, and Society.

Mienye, I. D., Sun, Y., & Ileberi, E. (2024). Artificial intelligence and sustainable development in Africa: A comprehensive review. *Machine Learning with Applications*, *18*, 100591. https://doi.org/10.1016/j.mlwa.2024.100591

Mill, J. S. (1861). *Utilitarianism* (2nd ed.). Hackett Publishing Co., Inc.

Mittelstadt, B. D., & Floridi L. (2016). The ethics of big data: Current and foreseeable issues in biomedical contexts. *Science and Engineering Ethics*, *22*(2), 303–341. https;//doi.org/10.1007/s11948-015-9652-2

Mubangizi, J. C. (2024). Artificial intelligence, human rights and sustainable development: An African perspective. *Perspectives of Law and Public Administration*, *13*(3), 374–389.

Muller, C. (2020). *The impact of artificial intelligence on human rights, democracy and the rule of law*. Council of Europe, Ad Hoc Committee on Artificial Intelligence (CAHAI).

Müller, V. C. (2020). Ethics of artificial intelligence and robotics. In E. N. Zalta (Ed.), *The Stanford encyclopaedia of philosophy*. Metaphysics Research Lab, Stanford University.

Næss, A. (1989). *Ecology, community and lifestyle: Outline of an ecosophyl*. Cambridge University Press.

Ndemo, B., & Thegeya, A. (2022). *Policy brief: A data governance framework for Africa*. African Economic Research Consortium. https://www.africaportal.org/documents/22658

Ndizera, V., & Muzee, H. (2018). A critical review of Agenda 2063: Business as usual?. *African Journal of Political Science and International Relations*, *12*(8), 142–154. https://doi.org/10.5897/AJPSIR2018.1114

Nemitz, P. (2018). Constitutional democracy and technology in the age of artificial intelligence. *Philosophical Transactions: Mathematical, Physical and Engineering Sciences*, *376*(2133), 20180089.

Nerurkar, M., Wadephul, C., & Wiegerling, K. (2016). Ethics of big data: Introduction. *International Review of Information Ethics*, *24*, 2–4. https://doi.org/10.29173/irie154

O'Neil, C. (2016). *Weapons of math destruction: How big data increases inequality and threatens democracy*. Penguin.

Obi, P. A. (2024). Labouring and smiling: Re-imagining digital colonialism in Africa: Silicon Valley big techs, and the politics of prosumer capitalism in Nigeria. *Triple c: Communication, Capitalism & Critique*, *22*(1), 381–395. https://doi.org/10.31269/triplec.v22i1.1451

Okolo, C. T., Aruleba, K., & Obaido, G. (2023). Responsible AI in Africa: Challenges and opportunities. In D. O. Eke, K. Wakunuma, & S. Akintoye (Eds.), *Responsible AI in Africa, social and cultural studies of robots and AI* (pp. 35–64). Palgrave Macmillan. https://doi.org/10.1007/978-3-031-08215-3_1

Omotubora, A., & Basu, S. (2024). Decoding and reimagining AI Governance beyond colonial shadows. In *Handbook on Public Policy and Artificial Intelligence* (pp.222–236). Edward Elgar.

Oti, P. A., & Otalor, J. I. (2019). Budgeting, governance and sustainable development in Nigeria: Military versus Democratic era. *International Journal on Governmental Financial Management*, *19*(2), 18–37.

Palomares, I., Martínez-Cámara, E., Montes, R., García-Moral, P., Chiachio, M., Chiachio, J., et al. (2021). A panoramic view and swot analysis of artificial intelligence for achieving the sustainable development goals by 2030: Progress and prospects. *Applied Intelligence: The International Journal of Artificial Intelligence, Neural Networks, and Complex Problem-Solving Technologies*, *51*, 6497–6527.

Patil, N. H., Patel, S. H., & Lawand, S. D. (2023). Research paper on artificial intelligence and it's applications. *Journal of Advanced Zoology*, *44*(S-8), 229–238.

Peralta, A., Olivas, J. A., Romero, F. P., & Navarro-Illana, P. (2025). Integration of fuzzy techniques and formal representation of domain and expert knowledge in AI systems: A comprehensive review. *Contemporary Mathematics*, *6*(2), 1660–1681. https://ojs.wiserpub.com/index.php/CM/article/view/6231

Potter, R. B., Binns, J. A., Elliott, J. A., & Smith, D. (2004). *Geographies of development* (2nd ed.). Addison Wesley Longman.

Pramanik, M. I., Lau, R. Y. K., Azad, M. A. K., Hossain, M. S., Chowdhury, M. K. H., & Karmaker B. K. (2020). Healthcare informatics and analytics in big data. *Expert Systems with Applications*, *152*, 113388. https://doi.org/10.1016/j.eswa.2020.113388

Rahman, A. U., Saqia, B., Alsenani, Y. S., & Ullah, I. (2024). Data quality, bias, and strategic challenges in reinforcement learning for healthcare: A survey. *International Journal of Data Informatics and Intelligent Computing*, *3*(3), 24–42. https://doi.org/10.59461/ijdiic.v3i3.128

Rainie, L., Anderson, J., & Vogels, E. A. (2021). *Experts doubt ethical AI design will be broadly adopted as the norm within the next decade*. Pew Research Centre.

Rasheed, I., Subhan, I., Ibrahim, M., & Sayed, F. A. (2015). Knowledge management as a strategy & competitive advantage: a strong influence to success, a survey of knowledge management case studies of different organizations. *Information and Knowledge Management*, *5*(8), 60–71.

Regilme, S. S. F. (2024). Artificial intelligence colonialism: environmental damage, labor exploitation, and human rights crises in the global south. *The SAIS Review of International Affairs*, *44*(2), 75–92. https://doi.org/10.1353/sais.2024.a950958

Richards, N. M., & King, J. H. (2014). Big data ethics. *Wake Forest Law Review*, *49*, 393–432.

Ross, W. D. (Trans.). (1999). *The Nicomachean ethics*. Batoche Books.

Ruggerio, C. A. (2021). Sustainability and sustainable development: A review of principles and definitions. *Science of The Total Environment*, *786*, 147481. https://doi.org/10.1016/j.scitotenv.2021.147481

Russell, S., & Norvig, P. (2020). *Artificial Intelligence: A modern approach* (4th ed.). Pearson.

Sætra, H. S. (2021). AI in context and the sustainable development goals: Factoring in the unsustainability of the sociotechnical system. *Sustainability*, *13*, 1–19, 1738. https://doi.org/10.3390/su13041738

Salami, A. O. (2024). Artificial intelligence, digital colonialism, and the implications for Africa's future development. *Data & Policy*, 6:e67. https://doi.org/10.1017/dap.2024.75

Sayed, F. A., & Muhammad, K. S. (2015). Impact of knowledge management on security of information. *Information and Knowledge Management*, *5*(1), 52–59.

Shaikh, T. A., Rasool, T., & Rasheed Lone, F. (2022). Towards leveraging the role of machine learning and artificial intelligence in precision agriculture and smart farming. *Computers and Electronics in Agriculture*, *198*, 107119. https://doi.org/10.1016/j.compag.2022.107119

Sheikh, H., Prins, C., & Schrijvers, E. (2023). Artificial intelligence: Definition and background. In H. Sheikh, C. Prins, & E. Schrijvers (Eds.), *Mission AI. Research for policy* (pp 15–41). Springer. https://doi.org/10.1007/978-3-031-21448-6_2

Siakwah, P., Musavengane, R., & Leonard, L. (2019). Tourism governance and attainment of the sustainable development goals in Africa. *Tourism Planning & Development*, *17*(4), 355–383. https://doi.org/10.1080/21568316.2019.1600160

Singh, A., Kanaujia, A., Singh, V. K., & Vinuesa, R. (2024). Artificial intelligence for sustainable development goals: Bibliometric patterns and concept evolution trajectories. *Sustainable Development*, *32*, 724–754.

Stahl, A. (2021a). How AI will impact the future of work and life. Retrieved December 12, 2025, from https://www.forbes.com/sites/ashleystahl/2021/03/10/how-ai-will-impact-the-future-of-work- and-life/

Stahl, B. C. (2021b). *Artificial intelligence for a better future*. Springer.

Stahl, B. C. (2025). The ethics of data and its governance: A discourse theoretical approach. *Information*, *16*(6), 497. https://doi.org/10.3390/info16060497

Stahl, B. C., & Eke, D. (2024). The ethics of ChatGPT – Exploring the ethical issues of an emerging technology. *International Journal of Information Management*, *74*, 102700. https://doi.org/10.1016/j.ijinfomgt.2023.102700

Stahl, B. C. (2012). Morality, ethics, and reflection: A categorization of normative IS research. *Journal of Association for Information Systems*, *13*(8), 636–656.

Stone, P., Brooks, R., Brynjolfsson, E., Calo, R., Etzioni, O., Hager, G., Hirschberg, J., Kalyanakrishnan, S., Kamar, E., & Kraus, S. (2016). *Artificial intelligence and life in 2030. One hundred year study on artificial intelligence: Report of the 2015–2016 study panel*. arXiv. arXiv:2211.06318.

Tapo, A. A., Traore, A., Danioko, S., & Tembine, H. (2024). *Machine intelligence in Africa: A survey*. ArXiv. https://doi.org/10.48550/arXiv.2402.02218

Tarhebai, C. E. (2024). *The Impact of artificial intelligence (ai) on the electrical systems in Buea–Cameroon* [Unpublished master's thesis]. Istanbul Gelisim University Institute of Graduate Studies.

Thore, S., & Tarverdyan, R. (2022). *Measuring sustainable development goals performance*. Elsevier. https://doi.org/10.1016/b978-0-323-90268-7.00002-5

Thusi, X., Mlambo, V. H., & Mkhize, N. E. (2024). Challenges of implementing the Africa Agenda 2063: A synopsis. *Prizren Social Science Journal, 8*(1), 58–66.

Tiwari, S. (2024). The rise of intelligent machines: An introduction to artificial intelligence. In A. Khanna, M. E. Barachi, S. Jain, M. Kumar, & A. Nayyar. (Eds.), *Artificial intelligence and machine learning in drug design and development* (pp. 1–22). Wiley. https://doi.org/10.1002/9781394234196.ch1

Tremblay, D., Fortier, F., Boucher, J. F., Riffon, O., & Villeneuve, C. (2020). Sustainable development goal interactions: An analysis based on the five pillars of the 2030 agenda. *Sustainable Development, 28*, 1584–1596.

Triguero, I., Molina, D., Poyatos, J., Del Ser, J., & Herrera, F. (2024). General Purpose Artificial Intelligence Systems (GPAIS): Properties, definition, taxonomy, societal implications and responsible governance. *Information Fusion, 103*, 102135. https://doi.org/10.1016/j.inffus.2023.102135

UNECA. (2017). *Assessing regional integration in Africa VIII: Bringing the Continental Free Trade Area about*. UNECA.

UNESCO. (2020). *First draft of the recommendation on the ethics of artificial intelligence*. UNESCO.

United Nations High Commissioner for Human Rights (UNHCHR). (2021). *The right to privacy in the digital age: Report of the United Nations High Commissioner for Human Rights*. CPIT. https://undocs.org/A/HRC/48/31

United Kingdom AI (UKAI) Council. (2021). *AI roadmap*. UK AI Council.

Valenzuela, A., Puntoni, S., Hoffman, D., Castelo, N., De Freitas, J., Dietvorst, B., Hildebrand, C., Huh, Y. E., Meyer, R., Sweeney, M. E., Talaifar, S., Tomaino, G., & Wertenbroch, K. (2024). How artificial intelligence constrains the human experience. *Journal of the Association for Consumer Research, 9*(3), 241–256. https://doi.org/10.1086/730709

Veale, M., Binns, R., & Edwards, L. (2018). Algorithms that remember: Model inversion attacks and data protection law. *Philosophical Transactions of the Royal Society, 376*(2133), 1–15. https://doi.org/10.1098/rsta.2018.0083

Verhoef, P. C., Broekhuizen, T., Bart, Y., Bhattacharya, A., Qi Dong, J., Fabian, N., & Haenlein, M. (2021). Digital transformation: A multidisciplinary reflection and research agenda. *Journal of Business Research, 122*, 889–901.

Vinuesa, R., Azizpour, H., Leite, I., Balaam, M., Dignum, V., Domisch, S., Felländer, A., Langhans, S. D., Tegmark, M., & Nerini, F. F. (2020). The role of artificial intelligence in achieving the Sustainable Development Goals. *Nature Communications, 11*(1), 233. https://doi.org/10.1038/s41467-019-14108-y

Willcocks, L. (2020). Robo-Apocalypse cancelled? Reframing the automation and future of work debate. *Journal of Information Technology, 35*, 286–302.

World Commission on Environment and Development. (1987). *Our common future*. Oxford.

Zuboff, S. (2019). *The age of surveillance Capitalism: The fight for a human future at the new frontier of power* (1st ed.). Profile Books.

Zwitter, A. (2014, July-December). Big data ethics. *Big Data and Society, 1*(2), 1–6. https://doi.org/10.1177/2053951714559253

Chapter 5
Sustainability Education in the Digital Age: Building a Future-Ready Workforce

Shahira Abdalla and Eseroghene Udjo

Abstract Higher education institutions worldwide are increasingly integrating digital technologies to enhance sustainability education, yet African universities continue to face systemic limitations that restrict meaningful transformation. Despite global commitments to the Sustainable Development Goals (SDGs) and the African Union Agenda 2063, significant gaps remain in technological capacity, faculty readiness, governance structures, and the development of future-ready competencies. This study examines how digital transformation is reshaping sustainability education in African higher education institutions and identifies the critical pillars needed to support effective and equitable implementation. A systematic literature review (SLR) was conducted using PRISMA guidelines, drawing on peer-reviewed studies published between 2017 and 2025 across major academic databases. Twenty-nine articles met the inclusion criteria and were analyzed using thematic synthesis aligned to five conceptual pillars: digital integration, competency development, faculty and institutional capacity, change management, and strategic governance. Findings show that African universities demonstrate increasing interest in digital sustainability initiatives; however, progress remains uneven due to infrastructural constraints, fragmented policy environments, and inconsistencies in faculty professional development. In contrast, institutions in Europe, Asia, and North America demonstrate more advanced, technology-enabled sustainability models. The study contributes a five-pillar conceptual framework that clarifies how digital transformation can support sustainability learning in resource-constrained contexts and offers policy, institutional, and pedagogical recommendations to guide future implementation and research.

Keywords Sustainability education · digital transformation · future-ready workforce · digital skills · competency skills · sustainable development goals

S. Abdalla · E. Udjo
Business Department, Higher Colleges of Technology, Ras Al-Khaimah, UAE
e-mail: sosama@hct.ac.ae

K. Modugu, J. Otalor (eds.), *Sustainability and Digital Transformation in Africa*, Frontiers in African Business Research,
https://doi.org/10.1007/978-981-95-9500-6_5

1 Introduction

The rapid pace of digital transformation is reshaping higher education globally, influencing instructional delivery, curriculum design, and the development of competencies required for a rapidly evolving workforce. At the same time, sustainability education has emerged as a critical priority for aligning higher education systems with international development agendas such as the United Nations Sustainable Development Goals (SDGs) and the African Union Agenda 2063. Although substantial research has examined the integration of digital tools in sustainability learning within European, Asian, and North American institutions, considerably less is known about how these dynamics unfold in African higher education systems. This lack of consolidated evidence limits the continent's ability to respond effectively to global sustainability demands and to leverage digital transformation for educational innovation.

Existing studies highlight fragmented implementation efforts across African universities, characterized by infrastructural constraints, limited digital governance frameworks, and uneven faculty capacity. However, the literature provides a limited synthesis of how these structural barriers intersect with sustainability education, and even fewer studies analyze the competencies required to prepare a future-ready African graduate. This reveals a significant research gap concerning the institutional, pedagogical, and governance conditions shaping digital sustainability education in Africa.

The purpose of this study is to examine how digital transformation is influencing sustainability education in African higher education institutions and to identify the competencies, institutional enablers, and strategic frameworks required to support this transition. To address this purpose, the study is guided by the following research questions:

1. How is digital transformation reshaping sustainability education in African higher education institutions?
2. What institutional, pedagogical, and governance factors influence the integration of digital transformation into sustainability education?
3. What digital and sustainability competencies are needed to prepare a future-ready African workforce?

Addressing these questions contributes to a deeper understanding of how African universities can advance digital sustainability initiatives despite systemic constraints and provides a foundation for policy, institutional innovation, and future research.

2 Literature Review and Theoretical Foundations

Digital transformation in higher education has evolved from a technological enhancement to a strategic and pedagogical imperative that reshapes how universities operate, teach, and engage with stakeholders. Scholars (Fernández et al., 2023; Zawacki-Richter, 2021) increasingly views digital transformation as a structural process that aligns institutional systems, governance, and culture toward new forms of learning and knowledge production. This transition is particularly significant for African higher education institutions, which must simultaneously navigate modernization pressures and entrenched structural challenges such as resource scarcity, limited infrastructure, and socioeconomic inequalities.

2.1 Global Landscape of Digital Transformation in Higher Education

International literature highlights a progression from early digital initiatives such as e-learning to more integrated digital ecosystems characterized by artificial intelligence, cloud technologies, virtual learning environments, and data driven decision-making. Research consistently shows that while digital tools offer opportunities for personalization, flexibility, and expanded access, their effectiveness depends on pedagogical alignment, institutional readiness, and cultural acceptance (Erkinovich & Feruz, 2025). Several scholars caution that universities often adopt technologies without rethinking underlying teaching philosophies, leading to superficial change rather than meaningful innovation. These debates underscore the complexity of digital transformation beyond technological acquisition.

2.2 Sustainability Education: Evolving Paradigms and Debates

Sustainability education has shifted from environmental awareness to a broader framework that incorporates systems thinking, futures literacy, socio-ecological responsibility, and transformative learning. Two major tensions appear in the literature. The first concerns whether sustainability education should emphasize measurable competencies or focus on deeper transformative learning involving values, identity, and worldview shifts. The second relates to the dominance of Western sustainability paradigms and the growing call for localized or Indigenous knowledge systems, particularly within the Global South. Another persistent challenge is the fragmentation of sustainability integration within curricula, where sustainability themes are often isolated rather than embedded across programs (Sterling, 2014; Lotz-Sisitka, 2016; Wamsler, 2020).

2.3 Intersections Between Digital Transformation and Sustainability Education

Studies at the intersection of digital transformation and sustainability education are emerging but remain limited (Wamsler, C. 2020; Krüger et al., 2025). Existing work suggests that digital platforms, virtual labs, simulation environments, and online communities can enhance systems thinking and collaborative problem solving, which are core to sustainability learning. For instance, countries in West Africa, particularly Nigeria and Ghana, have witnessed the volatile expansion of mobile learning platforms. However, the lack of consistent, strong governance and limitations of power supply can represent obstacles to optimum utilization of digital technologies (Rarhoui, 2024).

Learning analytics and artificial intelligence also present possibilities for measuring the development of sustainability competencies (Ifenthaler & Yau, 2020; Krüger, et al., 2025). However, critics argue that digital transformation can reproduce structural inequalities and prioritize technological efficiency over contextually grounded sustainability learning (Lotz-Sisitka, 2016; Wamsler, 2020). These tensions illustrate that digital technologies may support sustainability goals but cannot replace the social, cultural, and community-based dimensions required for meaningful sustainability education.

2.4 African Higher Education: Structural Realities and Capacity Gaps

The African higher education landscape is shaped by infrastructural constraints, funding limitations, and governance challenges that influence both digital transformation and sustainability education. In line with that, Congo and Chad are one of those African countries that are identified as digitally backward due to poor institutional governance and fragile telecommunications networks (Vhumbunu et al., 2023). Conversely, in South Africa, higher education institutions have strong governance and have well-established university sustainability centers with digital platforms (Vaye, 2025).

Many institutions face persistent digital divides between urban and rural campuses, limited autonomy for curriculum redesign, and rapidly shifting labor market demands that outpace curriculum reform (Teferra, D. 2017; Ashraf & Ademola, 2022). Scholars such as Rarhoui, 2024 and Vhumbunu et al. 2023, reported that initiatives related to sustainability and digital innovation frequently rely on donor funding, leading to fragmented, short-term projects instead of comprehensive systemic change. There is also an increasing body of work calling for decolonized sustainability education grounded in African epistemologies, community realities, and socio-ecological contexts (Lotz-Sisitka, 2016; Le Grange, 2016).

2.5 Digital Competencies, Sustainability Skills, and Labor Market Expectations

Labor markets across Africa are experiencing rapid change due to automation, climate transitions, and the expanding digital economy. Employers increasingly expect graduates to possess both digital literacy and sustainability awareness (Ahmed,

2025; Krüger, et al., 2025). As a result, research has focused on how universities can align their programs with emerging green and digital skills requirements (Ashraf et al., 2022). Furthermore, Van der Merwe et al., (2023) emphasized that while micro-credentials and short courses are gaining prominence, there is a lack of empirical studies examining how digital pedagogies contribute to the development of sustainability competencies in African higher education. Other studies such as Le Grange (2016) and Teferra (2017) highlighted significant mismatches between university training and workplace expectations.

2.6 Governance and Strategic Alignment

Governance is widely recognized as a critical determinant of successful digital and sustainability integration (European Commission, 2020). Internationally, institutions with formal digital strategies, sustainability frameworks, and cross-functional leadership structures are more likely to achieve coherent implementation. In African contexts, both Kenya and Rwanda have implemented comprehensive national digital education initiatives, supported by ICT hubs and robust partnerships between public and private institutions. It is worth noting that the Smart Education Master Plan of Rwanda is acknowledged as a model for regional digital policy alignment; however, governance structures often remain fragmented, and policy frameworks do not always translate into operational strategies (Materu, 2017). Funding limitations, regulatory constraints, and inconsistent institutional autonomy hinder long-term planning (Rarhoui, 2024). Where governance mechanisms such as sustainability offices or digital transformation roadmaps exist in Rwanda, they tend to enhance coordination and institutional focus, but such examples remain uneven across the continent (Sangwa, et al., 2025).

While the literature review provides an overview of broad trends in African higher education, it is important to recognize the persistent regional disparities that exist across the continent. Pathways toward digital sustainability transformation vary significantly due to differences in institutional governance, political stability, funding capacity, and institutional readiness. These factors collectively shape the ability of institutions to adopt and effectively implement digital and sustainability initiatives.

3 Conceptual Framework

This study is guided by a conceptual framework that explains how digital transformation supports sustainability education in African higher education institutions. The framework is organized as a structured sequence that moves from inputs, through enablers and processes, to outcomes. This structure clarifies how foundational conditions and institutional mechanisms interact to produce sustainability-oriented educational outcomes (Zawacki-Richter, 2021; Brooks & McCormack, 2020).

Rather than treating digital transformation and sustainability education as isolated initiatives, the framework integrates structural conditions, institutional capabilities, and strategic actions within a single explanatory model. Figure 5.1 presents this logic visually and illustrates how higher education institutions can progress toward future-ready, sustainability-aligned education (Trevisan et al., 2024).

At its core, the framework is built on five interrelated pillars: Digital Integration, Faculty and Institutional Capacity, Strategic Governance, Change Management, and Competency Development. These pillars are analytically distinct but operationally interconnected. For conceptual clarity, they are organized within four functional layers in the framework.

Digital Integration and Change Management are positioned within the process layer, as they represent the mechanisms through which institutions operationalize digital transformation (Henderson et al., 2017). Faculty and Institutional Capacity and Strategic Governance function as enablers, shaping the ability of institutions to translate infrastructure and policy conditions into effective action (European Commission, 2020; African Union, 2022). Competency Development represents the primary outcome, capturing the human-centerd objective of sustainability education (Sterling, 2014; Barth & Rieckmann, 2016). The input layer reflects the contextual foundations, including digital infrastructure, policy environment, and institutional resources, upon which all five pillars depend (UNESCO, 2024).

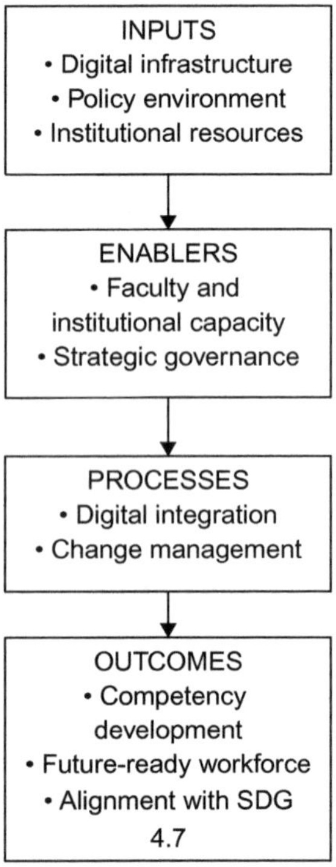

Fig. 5.1 Conceptual Framework for Digital Sustainability Education in African Higher Education

The framework therefore retains five conceptual pillars while presenting them within a coherent, layered structure. Alignment across these pillars determines whether digital transformation contributes meaningfully to sustainability education or results in fragmented, technology-driven implementation, a concern widely noted in studies of higher education digitalization in the Global South (Czerniewicz, 2018; Zawacki-Richter, 2021).

3.1 Inputs: Digital and Policy Foundations

The framework begins with foundational inputs. These include digital infrastructure, policy environment, and institutional resources. Together, they define the baseline conditions under which digital sustainability initiatives can emerge (Inter-American Development Bank, 2021; UNESCO, 2024). Digital infrastructure refers to connectivity, learning platforms, and access to devices that support teaching and administration. Policy environment captures national digital strategies, higher education regulations, and sustainability commitments. Institutional resources include funding, technical support, and organizational readiness. In many African contexts, limitations at this input stage constrain the scale and depth of digital transformation, even where institutional intent exists (Materu, 2017; African Union, 2022).

3.2 Enablers: Faculty and Institutional Capacity and Strategic Governance

Inputs alone are insufficient to drive transformation. The framework therefore positions faculty and institutional capacity and strategic governance as key enablers. Faculty and institutional capacity reflect the ability of educators and institutions to translate digital access into pedagogically meaningful practices. This includes faculty digital competence, curriculum innovation capability, and internal support systems. Without sustained professional development, digital tools remain underutilized (Zhao et al., 2021; Fernández et al., 2023). Strategic governance provides leadership direction and policy coherence. It includes institutional strategies, decision-making structures, accountability mechanisms, and alignment with national and international agendas. Where governance is fragmented or short-term, digital sustainability initiatives tend to be donor-driven and difficult to sustain (European Commission, 2020; African Union, 2022).

3.3 Processes: Digital Integration and Change Management

The process layer captures how institutions operationalize transformation. Two mechanisms are central: digital integration and change management. Digital integration refers to the purposeful embedding of digital technologies into teaching, learning, assessment, and administration. It moves beyond technology adoption to focus on alignment with sustainability learning goals. In African higher education, this process is often uneven due to infrastructural and capacity constraints (Inter-American Development Bank, 2021; Hanafi et al., 2025). Change management reflects an institution's ability to adapt. It includes curriculum redesign, industry engagement, partnership development, and responsiveness to labor market change. Effective change management enables institutions to contextualize global sustainability frameworks within local realities (African Union, 2022). Weak change management, by contrast, leads to slow reform and limited impact (Zawacki-Richter, 2021).

3.4 Outcomes: Competency Development and Future-Ready Graduates

Competency development results from aligned inputs, enablers, and processes. This outcome represents the human-centered objective of the framework. Competency extends beyond technical digital skills. They include systems thinking, ethical reasoning, critical judgement, collaboration, and sustainability awareness. These competencies align directly with Sustainable Development Goal 4.7, which emphasizes education for sustainable development and global citizenship (United Nations, 2015; UNESCO, 2021). The framework therefore positions the future-ready workforce as the ultimate outcome. Graduates emerge not only digitally literate but also capable of addressing complex sustainability challenges in digitally mediated environments (Redman & Redman, 2017; Krüger et al., 2025).

3.5 Integrative Role of Strategic Governance

Strategic governance functions as the anchoring element across the entire framework. It ensures coherence between inputs, enablers, processes, and outcomes. Strong governance aligns infrastructure investment with faculty development, supports change management, and embeds sustainability competencies within institutional missions (European Commission, 2020; UNESCO, 2024).

In regions such as Europe, coherent digital education policies provide continuity and funding stability. In contrast, many African higher education institutions operate without integrated governance frameworks, limiting long-term impact. This reinforces the framework's emphasis on governance as a critical condition for sustainable digital transformation.

4 Methodology

4.1 Review Design

This study uses a systematic literature review (SLR) to synthesize evidence on how digital transformation is shaping sustainability education in African higher education institutions (HEIs). The review follows PRISMA 2020 reporting guidance and applies a transparent, replicable search, screening, and appraisal workflow. The review focuses on peer-reviewed journal articles published from January 2017 to June 2025. The unit of analysis is the study. Policy documents and institutional reports were used only for background framing and were not counted as included studies unless they met the inclusion criteria.

4.2 Search Strategy

Searches were conducted in Scopus, Web of Science, ERIC, Google Scholar, and African Journals Online (AJOL). Keywords were developed from three concept blocks: (1) sustainability education, (2) digital transformation and digital pedagogy, and (3) higher education and Africa. Boolean operators and truncation were adapted to each database.

Table 5.1 reports the final PRISMA compliant search strategy and the number of records retrieved per database.

4.3 Eligibility Criteria

Studies were included if they: (a) examined sustainability education and or education for sustainable development in HEIs, (b) addressed digital transformation or digitally enabled teaching and learning, (c) reported findings relevant to Africa or provided transferable evidence for African HEIs, (d) were peer reviewed journal articles, and (e) were available in English.

Studies were excluded if they: (a) focused on primary or secondary education, (b) were conference papers, dissertations, editorials, or opinion pieces, (c) did not address either digital transformation or sustainability education substantively, or (d) lacked sufficient methodological detail for appraisal.

Table 5.1 PRISMA Search Strategy[a]

Author(s), Year	Context (Region)	Study Type	Primary Focus	Mapped Pillar(s)
Henderson et al., 2017	Global	Qualitative	Digital pedagogy	Digital Integration; Faculty Capacity
Materu, 2017	Sub-Saharan Africa	Policy review	Quality assurance reform	Strategic Governance; Change Management
Czerniewicz, 2018	South Africa	Qualitative	Digital inequality	Digital Integration; Strategic Governance
Zawacki-Richter, 2021	Global	Conceptual	Transformation complexity	Change Management; Strategic Governance
Ifenthaler & Yau, 2020	Global	Systematic review	Learning analytics	Competency Development; Digital Integration
Fernández et al., 2023	Global	Multivocal review	Digital initiatives	Faculty Capacity; Strategic Governance
Trevisan et al., 2024	Global	Systematic review	Digital sustainability	Strategic Governance; Change Management
Hanafi et al., 2025	Africa	Empirical	Digital transformation	Digital Integration; Strategic Governance
Ebom-Jebose, 2025	Africa	Empirical	Digital skills for SD	Competency Development; Digital Integration
Krüger et al., 2025	Global	Empirical	Sustainability competencies	Competency Development; Strategic Governance
Arregi et al., 2025	Europe	Empirical	Industry 5.0 skills	Faculty Capacity; Competency Development
García-Peñalvo, 2021	Global	Conceptual	Crisis-driven digitalization	Change Management; Strategic Governance
African Union, 2022	Africa	Policy	Digital education strategy	Strategic Governance; Digital Integration
UNESCO, 2021	Global	Policy review	Digital education for SD	Strategic Governance; Competency Development

Table 5.1 (continued)

Author(s), Year	Context (Region)	Study Type	Primary Focus	Mapped Pillar(s)
UNESCO, 2024	Global	Report	Technology and equity	Digital Integration; Strategic Governance
Sá & Cavalcante, 2020	Brazil	Empirical	Digital media in ESD	Digital Integration; Competency Development
Holmes et al., 2021	Global	Review	AI in education	Competency Development; Digital Integration
Brooks & McCormack, 2020	Global	Review	Institutional transformation	Strategic Governance; Change Management
Fuchs et al., 2020	Global	Review	Leadership for sustainability	Strategic Governance; Faculty Capacity
Shava & Manyike, 2022	Africa	Qualitative	Indigenous knowledge	Competency Development; Change Management
Lotz-Sisitka, 2016	Africa	Conceptual	Decolonized sustainability	Competency Development; Change Management
Redman & Redman, 2017	Global	Review	Sustainability skills	Competency Development
Barth & Rieckmann, 2016	Global	Conceptual	Competency frameworks	Competency Development
Wamsler, 2020	Global	Review	Transformative learning	Competency Development; Change Management
European Commission, 2020	Europe	Policy	Digital education policy	Strategic Governance
UN SDSN, 2022	Global	Report	Youth SDG action	Competency Development
Ahmed, 2025	Global	Conceptual	Human–AI skills	Competency Development
Henderson et al., 2017	Global	Qualitative	Student digital practices	Faculty Capacity; Change Management

Source: Authors' synthesis based on systematic literature review.
[a]Table 5.1 summarizes the characteristics of the studies included in the review.

4.4 Screening Procedure and PRISMA Flow

All records were exported to a reference manager and duplicates were removed before screening. Screening was completed in three stages: title screening, abstract screening, and full text eligibility assessment. In total, 782 records were identified. After removal of duplicates, 568 records were screened. A further 494 records were excluded at title and abstract stage. Seventy-four full texts were assessed for eligibility. Twenty-nine studies met all criteria and were included in the synthesis. The PRISMA flow is presented in Fig. 5.2.

4.5 Quality Appraisal Using MMAT

Methodological quality of included empirical studies was assessed using the Mixed Methods Appraisal Tool (MMAT). The MMAT was applied by matching each study to the appropriate design category and then scoring against the relevant five criteria for that category.

Appraisal results were used to weight interpretation during synthesis. Studies were not excluded solely based on MMAT score. Lower scoring studies were treated

Fig. 5.2 PRISMA Flow Diagram

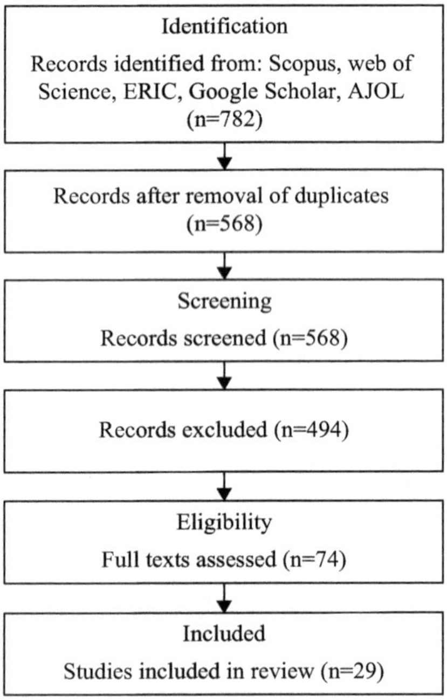

as providing indicative evidence rather than definitive conclusions. Table 5.2 summarizes MMAT outcomes by study type.

4.6 Data Extraction and Thematic Synthesis

A structured extraction form captured: bibliographic details, country or region, institutional context, study design, digital technology focus, sustainability education focus, and key findings. Extracted evidence was synthesized using thematic synthesis. Initial coding was inductive. Codes were then clustered into higher-order categories. Finally, categories were mapped to the conceptual framework to produce an integrated explanation of how digital transformation enables or constrains sustainability education in African HEIs.

4.7 Derivation of the Five Pillars

The five pillars underpinning the conceptual framework were derived through a systematic and iterative analytic process that combined inductive coding with conceptual consolidation, consistent with best practice in qualitative synthesis.

First, all included studies were subjected to open coding to identify recurrent drivers, constraints, enablers, and outcomes associated with digital transformation and sustainability education in higher education. This ensured sensitivity to both institutional and pedagogical dimensions. Second, initial codes were grouped into

Table 5.2 MMAT Quality Appraisal Summary by Study Type

Study type	n	MMAT criteria applied	Median criteria met	Interpretive note
Qualitative	9	Qualitative checklist (five items)	4/5 (80%)	Sampling clarity and reflexivity were the most common limitations.
Quantitative	6	Quantitative checklist (five items)	3.5/5 (70%)	Limitations related to measurement validity and control of confounders were common.
Mixed methods	5	Mixed methods checklist (five items)	4/5 (80%)	Integration across strands was the key differentiator of quality.
Review or multivocal review	6	Adapted transparency appraisal	4/5 (80%)	Quality varied based on reporting of search and synthesis steps.
Conceptual or policy sources	3	Not scored with MMAT	N/A	Used for framing and triangulation rather than evidence weighting.

higher-order categories capturing recurring patterns related to structural conditions, capability requirements, and implementation dynamics. These categories reflected common challenges and mechanisms observed across diverse institutional contexts. Third, an axial consolidation process was applied to examine conceptual overlap and interrelationships among categories. Categories were merged where they represented the same underlying construct and retained as distinct where they reflected different levers of institutional change. Fourth, the consolidated categories were formalized into five analytically distinct pillars: Digital Integration, Competency Development, Faculty and Institutional Capacity, Change Management, and Strategic Governance. Pillar labels were selected to balance conceptual precision with practical relevance, ensuring interpretability for both scholarly and policy-oriented audiences. Finally, the pillar structure was validated through cross-checking against the full set of included studies. Instances where empirical evidence did not align neatly with existing pillar definitions prompted refinement of conceptual boundaries rather than forced categorization.

These five pillars are operationalized and positioned within the layered structure of Fig. 5.1, which visualizes how evidence-derived constructs interact across inputs, enablers, processes, and outcomes.

5 Analysis

The systematic literature review of 29 studies revealed consistent global recognition of digital transformation as a central driver of innovation in higher education, particularly in reshaping curriculum delivery, assessment practices, and learner engagement. Across regions such as Latin America, the Caribbean, and Europe, universities acknowledge digital learning as essential for future competitiveness; however, substantial maturity gaps were identified. For example, the Inter-American Development Bank (2021) reported that although a majority of institutional leaders emphasized the importance of digital transformation, only a quarter assessed their institutions as highly digitally mature. This indicates that enthusiasm for digital adoption is not necessarily matched by institutional readiness or infrastructural capacity.

In African higher education institutions, the review identified widespread infrastructural limitations including poor broadband connectivity, high data costs, limited access to digital devices, and outdated technological systems (Zawacki-Richter, 2021; Vhumbunu et al., 2023; Rarhoui, 2024). These constraints hinder the integration of advanced technologies such as artificial intelligence, simulation-based laboratories, data analytics tools, and virtual learning systems, all of which are increasingly common in digitally advanced institutions globally. As a result, African universities tend to implement basic digital tools rather than pedagogically rich and innovation-oriented systems.

Across the studies, a strong emphasis was placed on competency development as a critical component of sustainability education. Future-oriented competencies such as digital literacy, systems thinking, ethical decision making, interdisciplinary problem solving, and entrepreneurial awareness were frequently highlighted as essential for preparing graduates for emerging green and digital labor markets (Redman & Redman, 2017; UNESCO, 2021; Krüger et al., 2025). Institutions such as Stanford University and Arizona State University were noted for their use of immersive, simulation-based learning approaches that foster sustainability competencies. In contrast, African universities tend to emphasize foundational digital skills due to limited access to advanced instructional technologies, resulting in uneven competency development and limited exposure to higher-level sustainability practices.

The analysis also revealed that faculty readiness plays a decisive role in shaping the success of digital sustainability education. Countries with structured and institutionalized faculty development programs, such as those in Scandinavia, demonstrate more consistent implementation of digital pedagogies. Conversely, African institutions often lack coherent professional development strategies, leading to fragmented adoption of digital teaching methods and inconsistent integration of sustainability content (Zhao et al., 2021; Fernández et al., 2023). Institutional capacity issues, including resource limitations, governance inconsistencies, and technical support shortages, further contribute to uneven implementation.

Change management emerged as another critical factor, with global evidence showing that strong industry academia partnerships enhance the relevance of sustainability programs and improve graduate employability outcomes. However, African universities face significant bureaucratic barriers that delay or prevent effective partnerships, program updates, and innovation adoption (Brooks & McCormack, 2020; Wamsler, 2020; Ashraf & Ademola, 2022). Additionally, there is limited longitudinal research assessing the long-term impact of digital and sustainability initiatives in African institutions, pointing to a major gap in the evidence base.

Finally, the results indicate that governance structures significantly influence the extent to which digital transformation and sustainability education can be integrated. European institutions benefit from comprehensive policy frameworks such as the EU Digital Education Action Plan and Green Deal initiatives, which provide strategic direction and funding mechanisms. In contrast, African universities often lack integrated policies and national level coordination, resulting in fragmented strategies, inconsistent implementation, and limited institutional accountability (European Commission, 2020; African Union, 2022; Sangwa et al., 2025).

6 Findings

The findings strongly support the five pillars (Digital Integration, Competency Development, Faculty and Institutional Capacity, Change Management, and Strategic Governance) of the conceptual framework developed for this study and highlight important implications for how digital transformation is reshaping sustainability education in African higher education systems. The persistent digital infrastructure gaps across African universities confirm that digital integration remains the most foundational challenge (Materu, 2017; UNESCO, 2024; Rarhoui, 2024). Without reliable connectivity, modern learning platforms, and access to digital devices, institutions cannot fully participate in global advances in digital sustainability education. The maturity gap seen in other regions is therefore even more pronounced in African systems, underscoring the need for targeted government and institutional investment.

The results also demonstrate that sustainability education increasingly depends on the development of complex, future-oriented competencies. Although global institutions are advancing rapidly in integrating simulation-based learning, virtual labs, and interdisciplinary digital tools, African universities remain constrained by limited technology access and insufficient training in digital pedagogies (Redman & Redman, 2017; Barth & Rieckmann, 2016; Krüger et al., 2025). This raises serious concerns about graduate preparation for emerging digital and green economies. The conceptual framework's emphasis on competency development as a key mediating factor is reinforced, as African students are at risk of being excluded from high-value skill sets required by modern industries.

Faculty readiness emerges as a central determinant of successful integration of digital sustainability curricula. The inconsistency of faculty development across African universities highlights a critical capacity gap, as educators are expected to adopt new pedagogies without adequate training or institutional support (Zhao et al., 2021; Fernández et al., 2023). This finding aligns with the framework's assertion that human capacity is as essential as technological capacity. Institutions that fail to invest in faculty development risk widening the gap between global standards and local realities.

The results further illustrate that institutional change processes in African universities are constrained by bureaucratic systems and rigid governance structures. These barriers prevent institutions from forming dynamic industry partnerships and adapting curricula to rapidly changing labor market demands (Zawacki-Richter, 2021; García-Peñalvo, 2021; Ashraf & Ademola, 2022). The limited availability of longitudinal research on African digital sustainability initiatives suggests that many reforms are short-lived or insufficiently evaluated, leaving institutions without evidence to inform future strategies. This validates the conceptual framework's positioning of change management as a catalyst for transformation.

Finally, the discussion highlights the critical role of governance in guiding digital sustainability reform. While regions such as Europe benefit from comprehensive digital and sustainability policies, African institutions often operate without unified strategies or adequate funding mechanisms. As a result, digital transformation and sustainability initiatives tend to be isolated, or donor-driven rather than embedded within long-term institutional visions (European Commission, 2020; African Union, 2022; Sangwa et al., 2025). This reinforces the framework's assertion that strategic governance acts as the overarching pillar connecting infrastructure, competencies, faculty capacity, and innovation processes.

Overall, the results and discussion reveal that digital transformation and sustainability education are deeply interconnected fields that must be addressed holistically. For African universities to advance meaningfully, they must move beyond fragmented initiatives toward integrated strategies that align infrastructure, pedagogy, capacity building, partnerships, and governance. The findings strongly support the view advanced by UNESCO (2021) that digital transformation and sustainability are mutually reinforcing and must be pursued together to realize equitable, future ready higher education systems in Africa.

7 Discussion

The findings validate the five pillars of the conceptual framework—Digital Integration, Competency Development, Faculty and Institutional Capacity, Change Management, and Strategic Governance, but also reveal critical tensions and contradictions that refine our understanding of how these pillars operate in African contexts. The digital integration pillar is confirmed as foundational, but findings show a misalignment between available technologies and pedagogical needs. Rather than seamless integration, most African institutions are improvising digital solutions (e.g., WhatsApp learning, mobile platforms), pointing to a "survivalist" model of digital transformation (Czerniewicz, 2018; Rarhoui, 2024). This highlights the need to adapt the pillar toward more context-responsive digital strategies rather than imported models of transformation.

The competency development pillar is crucial for adapting to the changing demands of the modern workforce. The findings strongly links with UNESCO's key competencies for sustainability, including systems thinking, normative competence, strategic competence, collaboration, critical thinking, self-awareness, and integrated problem-solving (UNESCO, 2021; Krüger et al., 2025). However, it is evident that the digital education programs at African universities examined in this study are limited on basic technical competencies with minimal integration of broader set of UNESCO-defined competencies essential for sustainability and digital transformation. Hence, this suggests an urgent need to reframe curricula not just around digital access, but around transformative competencies aligned with SDG 4.7.

The pillar of faculty and institutional capacity functions as both enabling factor and a bottleneck. The framework depends on established systems which support capacity development, yet organizations face challenges because they lack proper training programs and resources even when their countries support them (Zhao et al., 2021; Fernández et al., 2023). The results demonstrate that having enough capacity does not guarantee success because organizations need to convert their resources into actual operational results.

Change management is frequently seen as a constraint by bureaucratic inertia and short policy windows, often resulting in short-lived or inadequately evaluated reforms (Zawacki-Richter, 2021; Trevisan et al., 2024). These limitations underscore the need for greater institutional resilience and adaptability to ensure that change initiatives have long-lasting and meaningful impact.

Strategic governance emerged as an integrative pillar. However there is misalignment between policy formulation and execution (Materu, 2017; African Union,

2022; Sangwa et al., 2025). The lack of sustainability mandates and insufficient funding at many institutions results in the development of isolated donor-funded projects which do not create sustained institutional transformation. While countries like Rwanda demonstrate best practices, these remain exceptions. This study calls for urgency of aligning governance systems with institutional mission, national strategic goals and international standards to promote robust digital and sustainability education.

The results validate SDG 4.7 target because it requires all students to learn about sustainable development through their education. The implementation gaps which exist in African higher education institutions indicates misalignment between worldwide agreements about education and local capacities. Without funding for governance and faculty development, reaching goal 4.7 will be symbolic rather than transformative.

SDG 8 (Decent Work and Economic Growth) is also relevant, especially given that graduates lacking digital and sustainability skills are less employable in emerging green and digital sectors (Ahmed, 2025; Krüger et al., 2025). The framework's outcome focuses on future-ready graduates aligning with this SDG but requires stronger integration of labor market intelligence into curriculum reform.

The African Union's Digital Transformation Strategy emphasizes "inclusive digital education ecosystems" and "digitally skilled citizens." The findings reveal partial alignment, particularly in initiatives like Rwanda's Smart Education Master Plan, but most institutions remain at pilot stage or rely on donor-driven, nonintegrated projects (African Union, 2022; Rarhoui, 2024).

This study reveals evidence supporting the implementation phase of the African Union Digital Transformation Strategy for Africa (2020–2030), especially under Priority Action Area 4 (Education and Human Capital Development). However, it also suggests the need for regional capacity development hubs, shared digital infrastructure, and cross-border knowledge networks to address structural inequities.

In conclusion, we cannot emphasize more the importance of aligning policy and practice. Stakeholders must foster context-sensitive digital integration strategies, embed UNESCO's competencies into learning outcomes, launch national digital sustainability training programs, and develop accountability frameworks linking higher education institutions to SDG indicators.

8 Conclusion and Recommendations

This chapter examined how digital transformation is reshaping sustainability education in African higher education institutions, with the purpose of identifying essential competencies, institutional enablers, and strategic frameworks required to prepare future-ready graduates. Through a systematic literature review, the study synthesized global and regional insights across five core thematic domains, namely digital integration, competency development, faculty and institutional capacity, change management, and strategic governance.

The findings revealed significant variation in the global adoption of digital sustainability education. Institutions in Europe, Asia, and North America are at the forefront, characterized by the integration of artificial intelligence, simulation laboratories, immersive platforms, and smart campus technologies that support experiential and data-driven sustainability learning. In contrast, African higher education institutions are in the early stages of digital sustainability implementation. Many rely on mobile based microlearning models, WhatsApp mentoring, community innovation hubs, and low bandwidth tools due to limited digital infrastructure, inconsistent capacity building programs, insufficient governance mechanisms, and restricted industry engagement.

A key insight from the findings is that the urgency placed on digital literacy development in African institutions, while necessary, may inadvertently limit attention to higher-level cognitive competencies required for sustainability education. These competencies include systems thinking, ethical reasoning, interdisciplinary problem solving, and green entrepreneurship. Current institutional approaches appear fragmented and lack coordinated policy support from governments and regulatory bodies. The study emphasizes the need for integrated digital and sustainability frameworks to guide strategic planning, curriculum reform, and governance practices across African higher education.

The chapter concludes that the transformation of sustainability education in Africa requires holistic alignment between infrastructure, pedagogy, institutional governance, human capital development, and national policy direction. Such alignment would enable institutions to respond effectively to emerging global challenges and produce graduates capable of contributing to sustainable development in digitally mediated environments.

Although this study provides a strategic synthesis of digital sustainability education across African and international contexts, several limitations must be acknowledged. First, the study relies exclusively on a systematic literature review of peer reviewed studies published between 2017 and 2025 and does not incorporate primary empirical data. Second, the review treats Africa as a unified educational context and does not account for regional variations in policy environments, economic resources, cultural dynamics, or institutional histories. This may limit the generalizability of certain findings across diverse African countries.

Additionally, the rapid evolution of digital technologies means that some EdTech tools and practices captured in the reviewed studies may become outdated shortly after publication. As a result, some of the insights should be interpreted as indicative rather than definitive. Finally, the study excludes gray literature, institutional reports, and unpublished data that might offer valuable insights into the operational realities of digital sustainability implementation in Africa. These limitations suggest that the present findings should be viewed as foundational and as a stepping stone for more detailed empirical and context-specific research.

Achieving meaningful transformation of sustainability education through digital technologies in African higher education requires coordinated action across policy, institutional, pedagogical, and multistakeholder domains. At the policy level, national governments should develop comprehensive digital sustainability education frameworks aligned with global development agendas including the Sustainable Development Goals and the African Union Agenda 2063. These frameworks should define core digital and sustainability competencies, guide curriculum development, and promote multi sector collaboration. Governments should further establish funding mechanisms such as innovation grants, tax incentives, and public private investment schemes to support the deployment of digital sustainability programs. Accreditation bodies are encouraged to revise their quality assurance criteria to position digital and sustainability competencies as essential elements of institutional evaluation.

At the institutional level, universities need to strengthen governance systems by embedding digital transformation and sustainability within their strategic identities. This may involve establishing Chief Digital and Sustainability Officer roles, creating interdisciplinary transformation task forces, and redesigning internal structures to support collaboration across faculties. Faculty development should be prioritized through sustained professional learning programs focused on digital pedagogy, interdisciplinary instruction, and sustainability related curriculum design. Institutions should also invest in open-source digital platforms that support mobile access and low bandwidth connectivity to ensure equitable learning environments for students and educators.

From a pedagogical standpoint, universities should adopt competency-based education models that integrate systems thinking, digital literacy, ethical reasoning, and green entrepreneurship as core learning outcomes. Experiential and problem-based learning approaches need to be embedded across programs to mirror real world sustainability challenges. Integrating local and Indigenous knowledge into digital formats would strengthen cultural relevance and enable students to approach sustainability challenges through contextually grounded perspectives. The combination of community innovation labs, simulation tools, and digital storytelling offers meaningful opportunities for immersive and participatory learning.

Strengthening industry and multi-stakeholder partnerships is also essential. Higher education institutions should work closely with employers, governmental agencies, technology firms, and civil society organizations to design academic programs that reflect the evolving needs of digital and green economies. University based entrepreneurial ecosystems, including sustainability focused incubators, innovation bootcamps, competitions, and seed funding initiatives, should be expanded to support student led innovation. Such partnerships ensure that sustainability education remains relevant, applied, and aligned with labor market demands.

Finally, further research is essential to deepen understanding of digital sustainability education in Africa. Empirical studies conducted within specific African countries would help clarify how national contexts shape institutional readiness and student outcomes. Longitudinal research is particularly valuable for assessing how digital sustainability education influences student career readiness, innovation capacity, and societal engagement over time. Additional research on inclusive digital pedagogy would also help identify how digital tools can expand participation among women, rural learners, and other marginalized groups.

References

African Union. (2022). Digital education strategy and implementation plan. https://unevoc.unesco.org/bilt/bridging-africa/1.-DES-EN-2022_09_14.pdf

Ahmed, S. A. (2025). Reimagining education in the coming decade: What AI reveals about what really matters. *Frontiers in Education*, 10. https://doi.org/10.3389/feduc.2025.1699106

Arregi Lopez, A., Igartua, J. I., Retegi, J., & Eguren, J. A. (2025). Addressing the challenges associated with Industry 5.0 in higher education: A training concept approach. *The TQM Journal*. https://doi.org/10.1108/TQM-12-2024-0524

Ashraf, M. A., & Ademola, B. (2022). Curriculum reform in Sub-Saharan Africa: Navigating mismatch with labour market expectations. *Compare: A Journal of Comparative and International Education*, *52*(3), 456–475. https://doi.org/10.1080/03057925.2020.1804994

Ashraf, M. A., Iqbal, J., Arif, M. I., & Asghar, M. Z. (2022). Fostering ICT competencies in blended learning: Role of curriculum content, material, and teaching strategies. *Frontiers in Psychology*, 13, 758016. https://doi.org/10.3389/fpsyg.2022.758016

Barth, M., & Rieckmann, M. (2016). State of the art in research on higher education for sustainable development. In M. Barth, G. Michelsen, M. Rieckmann, & I. Thomas (Eds.), *Routledge handbook of higher education for sustainable development* (pp. 100–113). Routledge.

Bates, A. W. (2019). *Teaching in a digital age: Guidelines for designing teaching and learning*. BCcampus.

Brooks, C. D., & McCormack, M. (2020). *Driving digital transformation in higher education*. ECAR Research Report.

Czerniewicz, L. (2018). *Inequality as higher education goes online*. Inside Higher Ed.

Ebom-Jebose, A. (2025). Enhancing digital skills education for sustainable development and unemployment reduction in the digital economy. *European Journal of Innovative Studies and Sustainability*, *1*(1), 24–30. https://doi.org/10.59324/ejiss.2025.1(1).04

Erkinovich, I. A., & Feruz, Q. (2025). *Modern approaches to enhancing students' knowledge through innovative pedagogical strategies*. In Proceedings of the London International Conference on Multidisciplinary Research.

European Commission. (2020). *Digital education action plan 2021–2027: Resetting education and training for the digital age*. https://ec.europa.eu

Fernández, A., Gómez, B., Binjaku, K., & Meçe, E. K. (2023). Digital transformation initiatives in higher education institutions: A multivocal literature review. *Education and Information Technologies*, *28*, 12351–12382.

Fuchs, C., et al. (2020). Sustainability in higher education institutions: Leadership and institutional commitment. *Journal of Cleaner Production*.

García-Peñalvo, F. J. (2021). Digital transformation in higher education: The COVID-19 edition. *Education in the Knowledge Society*, *22*, Article e23623.

Hanafi, M., Kably, H., Lali, K., Elkouche, A., & Mahrez, F. (2025). Investigating the digital transformation of higher education in Africa. *EDP Audit, Control, and Security Newsletter*. https://doi.org/10.1080/07366981.2025.2538254

Henderson, M., Selwyn, N., & Aston, R. (2017). What works and why? Student perceptions of 'useful' digital technology in university teaching and learning. *Studies in Higher Education*, *42*(8), 1567–1579.

Holmes, W., Bialik, M., & Fadel, C. (2021). *Artificial intelligence in education: Promises and implications for teaching and learning*. Center for Curriculum Redesign.

Ifenthaler, D., & Yau, J. Y. K. (2020). Utilising learning analytics for study success: Reflections on current empirical findings. *Asian Journal of Distance Education*, *15*(2), 79–99.

Inter-American Development Bank. (2021). *Higher education digital transformation in Latin America and the Caribbean*. https://doi.org/10.18235/0003829

Krüger, C., Pennabel, A. F., & de Oliveira, M. S. B. (2025). Model for fostering competencies for sustainable development in HEI: Human resources as drivers of the process. *International Journal of Management Education*, *2*(3). https://doi.org/10.1016/j.ijme.2025.101238

Le Grange, L. (2016). Decolonising the university curriculum. *South African Journal of Higher Education*, *30*(2), 1–12. https://doi.org/10.20853/30-2-709

Lotz-Sisitka, H. (2016). Decolonizing knowledge for sustainability education in Africa. *International Review of Education, 62*, 343–364.

Materu, P. (2017). *Higher education quality assurance in Sub-Saharan Africa: Status, challenges, opportunities*. World Bank.

Page, M. J., et al. (2021). The PRISMA 2020 Statement: An updated guideline for reporting systematic reviews. *Systematic Reviews, 10*(89). https://doi.org/10.1186/s13643-021-01626-4

Redman, C. L., & Redman, S. (2017). Which sustainability competencies should be developed?. *Sustainability: The Journal of Record, 10*(6), 343–350.

Rarhoui, K. (2024). An analysis of African E-Government development: Trends and challenges. *African Scientific Journal, 10*(2). https://www.afrsj.com/index.php/AfricanScientificJournal/article/view/654

Sá, J., & Cavalcante, P. (2020). Sustainability education and digital media in Brazilian universities. *Latin American Education Journal, 29*(3), 54–68. https://doi.org/10.1108/IJSHE-07-2020-0247

Sangwa, D. S., Ruhamya, P., & Ekosse, E. (2025). Bridging gaps in Rwanda's higher education: A strategic analysis of program availability, delivery models, and regional competitiveness. *SSRN, 13*(2). https://doi.org/10.2139/ssrn.5283935

Shahzad, S. A., Pervaiz, Z., & Naqvi, S. A. J. (2025). Harnessing AI for promoting sustainability and life skills in 21st-century education. *Research Journal for Social Advancement, 3*(5), 679–688. https://rjsaonline.com/journals/index.php/rjsa/article/view/367

Shava, S., & Manyike, T. (2022). Indigenous knowledge and sustainability learning in African universities. *African Journal of Environmental Education, 18*(1), 1–18.

Sterling, S. (2014). *Sustainable education: Re-visioning learning and change*. Green Books.

Teferra, D. (2017). African higher education: The dynamics of international partnerships and donor influence. *International Journal of African Higher Education, 4*(2), 5–22. https://doi.org/10.6017/ijahe.v4i2.10256

Trevisan, L. V., Eustachio, J. H. P. P., & Dias, B. G. (2024). Digital transformation towards sustainability in higher education: State of the art and future research insights. *Environment, Development and Sustainability, 26*, 2789–2810. https://doi.org/10.1007/s10668-022-02874-7

UNESCO. (2024). Technology in education: A tool on our terms. 2024 Youth Report. https://www.unesco.org/gem-report/en/publication/2024-technology-education-tool-our-terms

UNESCO. (2021). *Digital education for sustainable development: A policy review*. UNESCO. https://unesdoc.unesco.org/ark:/48223/pf0000374309

UN Sustainable Development Solutions Network. (2022). *Youth Solutions Report 2022*. https://www.local2030.org/library/view/269

United Nations. (2015). *Transforming our world: The 2030 Agenda for Sustainable Development*.

Van der Merwe, T. M. D., Serote, M., & Maloma, M. (2023). A systematic review of the challenges of e-learning implementation in Sub-Saharan African countries (2016–2022). *Electronic Journal of e-Learning, 21*(5), 415–430. https://doi.org/10.34190/ejel.21.5.3075

Vaye, C. N. (2025). From national to continental priority: The recent trend of information communication technologyin Sub-Saharan African higher education. In *Digital Transformation and Higher Education in Africa*. Edward Elgar Publishing.

Vhumbunu, C. H., Adetiba, T. C., & Vyas-Doorgapersad, S. (2023). Digital transformation in higher education institutions in Africa: Challenges, opportunities, and policy implications. *International Journal of Educational Development*, 98, 102726. https://doi.org/10.1016/j.ijedudev.2023.102726

Wamsler, C. (2020). Education for sustainability: A transformative learning perspective. *Journal of Cleaner Production, 249*, 119–142.

Yadav, R., & Yadav, A. (2025). Effect of flipped classroom instruction on academic performance among future teacher educators. *International Journal of Education and Management Studies, 15*(3), 373–379.

Zawacki-Richter, O. (2021). Digital transformation in higher education: The complexity of change. *International Journal of Educational Technology in Higher Education, 18*, 1–6.

Zhao, Y., Llorente, A. M. P., & Sánchez Gómez, M. C. (2021). Digital competence in higher education research: A systematic literature review. *Computers & Education, 168*, 104212.

Chapter 6
Digital Communication Practices and Sustainability Messaging in African Contexts: A Cultural Linguistics and Pragmatic Perspective

Dare Eriel Ehigie and Jimoh Junior Braimoh

Abstract As African countries undergo rapid digital transformation, sustainability campaigns increasingly rely on mobile and social media platforms to reach diverse populations. However, the effectiveness of these digital interventions depends not only on technological access but on the pragmatic and cultural appropriateness of the messages conveyed. This chapter examines how sustainability messaging operates across multilingual African contexts, specifically in Nigeria, Kenya, and Ghana, drawing on theories from pragmatics, intercultural communication, and cultural linguistics. Through a grounded analysis of digital messages delivered via WhatsApp, Facebook, SMS, and mobile applications, the chapter explores how speech acts, politeness strategies, metaphor use, and narrative framing affect public interpretation and engagement. The study identifies key areas where pragmatic failures, such as overly direct speech or culturally inappropriate metaphors, have led to misinterpretation or resistance. Conversely, it highlights the communicative power of culturally anchored strategies, including the use of proverbs, storytelling, and identity-based appeals. Audience responses are analyzed to understand how digital messages are interpreted cognitively, emotionally, and behaviorally within local semiotic frameworks. The chapter argues that effective sustainability communication in Africa must transcend mere translation to engage local communicative norms, ethical codes, and symbolic repertoires. It calls for the integration of linguists, cultural mediators, and community leaders into message design and advocates for participatory, inclusive models that reflect local epistemologies. By advancing a linguistically and culturally responsive framework, the study contributes new insights into the human dimensions of digital sustainability in Africa.

D. E. Ehigie
State University of New York at Buffalo, USA

J. J. Braimoh
University of Mississippi, USA
e-mail: jimoh2braimoh@gmail.com

K. Modugu, J. Otalor (eds.), *Sustainability and Digital Transformation in Africa*, Frontiers in African Business Research,
https://doi.org/10.1007/978-981-95-9500-6_6

Keywords Pragmatics · intercultural communication · digital sustainability · African multilingualism · cultural linguistics · audience interpretation

1 Introduction

The advent of digital communication technologies has redefined how sustainability messages are conceptualized, delivered, and interpreted across African societies. With platforms like WhatsApp, Facebook, SMS broadcasts, and mobile applications gaining prominence, particularly in sub-Saharan Africa, digital channels now serve as critical vehicles for public engagement and behavioral change (Ajani et al., 2024; Keet, 2022). While these platforms offer speed and broad reach, the effectiveness of the messages they carry is deeply contingent upon their cultural and linguistic relevance, especially within linguistically diverse and multicultural environments (Ngoaketsi et al., 2024; Ehigie & Braimoh, 2024). Communication practices in many African settings are deeply rooted in indigenous epistemologies and sociolinguistic traditions such as communal storytelling, proverb usage, politeness norms, and layered indirectness (Braimoh, 2024; Igbinovia et al., 2024). However, sustainability messages distributed digitally often replicate Western communication patterns, characterized by direct speech acts, decontextualized imagery, and unfamiliar metaphors, that may lead to misinterpretation, resistance, or disengagement (Onomejoh et al., 2024; van Stam, 2021). The gap between digitally encoded intent and culturally mediated interpretation highlights a pressing need for context-sensitive frameworks that incorporate the sociopragmatic and cultural-linguistic ecologies of African audiences (Adebara & Abdul-Mageed, 2022; Oyekunle, 2025).

Emerging literature in applied linguistics, pragmatics, and intercultural communication has demonstrated the pivotal role of cultural conceptualizations in shaping message reception and interpretation (Sharifian, as discussed in Ajani et al., 2024; Foth & Hearn, 2007). Speech act theory and politeness strategies are central to the effectiveness of digital directives, warnings, or calls to action (Braimoh & Ehigie, 2024; Eke et al., 2025). When these acts are poorly calibrated to local norms, particularly in multilingual contexts like Nigeria, Ghana, or Kenya, the intended effects of digital sustainability campaigns are often undermined (Bhanye & Maisiri, 2023; Issaka et al., 2024). Furthermore, with increasing digital inclusion, marginalized linguistic communities are being brought into sustainability discourses via platforms they previously lacked access to. This necessitates not only multilingual translation but also cognitive alignment with locally embedded meaning systems (Mbogho et al., 2025; Adebara & Abdul-Mageed, 2022). Consequently, campaigns that integrate indigenous communicative modes, such as metaphoric expression, communal referencing, and culturally familiar symbols, tend to exhibit stronger public engagement and higher adoption of sustainable practices (Ehigie & Braimoh, 2024; Awuor, 2024).

Despite this awareness, a significant gap persists in how digital sustainability messaging is designed and evaluated. Many campaigns still overlook linguistic diversity and the cultural semiotics that structure African cognition (Keet, 2022; El Louadi, 2024). Scholars such as Litre et al. (2022) and UNESCO (2025) emphasize the urgency of embedding local linguistic and cultural frameworks into sustainability communication to prevent digital interventions from becoming culturally alienating or ineffective. This chapter positions itself at the intersection of pragmatics, cultural linguistics, and digital communication studies, examining how sustainability messages can be pragmatically and culturally optimized for multilingual African settings. Drawing on empirical insights and theoretical contributions from Nigeria, Kenya, and Ghana, this chapter examines digital messages in various domains, including environmental conservation, climate action, and public health. It applies analytical tools from speech act theory, politeness theory, and cultural linguistics to unpack how communicative strategies affect reception and behavioral outcomes (Ehigie & Braimoh, 2024; Onomejoh et al., 2024; Oyekunle, 2025).

The following four objectives guide the chapter:

- To explore how pragmatic features, particularly speech acts and politeness strategies, shape the effectiveness of digital sustainability messages in African contexts.
- To examine the role of indigenous communicative practices, such as proverbs, storytelling, and metaphor, in enhancing message salience and interpretation.
- To analyze the cultural and linguistic factors contributing to pragmatic failures or communication breakdown in digital sustainability campaigns.
- To propose a culturally responsive framework for designing digital sustainability messages that align with local interpretive ecologies.

2 Literature Review

2.1 Pragmatics and Speech Act Theory in Digital Communication

The application of speech act theory to digital discourse has provided scholars with a robust framework for understanding how language functions to achieve specific communicative goals. Speech acts such as directives, requests, and warnings form the core structure of sustainability messages and influence how audiences respond to calls for behavioral change (Braimoh & Ehigie, 2024; Eke et al., 2025). Within African digital environments, these acts are often filtered through culturally embedded norms concerning authority, politeness, and community expectation. The structure and illocutionary force of a message can either facilitate or hinder audience engagement depending on the level of alignment with local pragmatic expectations (Ehigie & Braimoh, 2024; Onomejoh et al., 2024). Researchers have observed that digital platforms, such as WhatsApp and Facebook, often amplify directness in speech acts, which may conflict with the culturally preferred indirectness and contextual nuance in African communities (Ngoaketsi et al., 2024; Ajani et al., 2024).

Therefore, messages that intend to inform or persuade may inadvertently offend or alienate audiences when they bypass sociopragmatic norms such as age-based deference or community-based approval mechanisms.

2.2 Cultural Linguistics and Indigenous African Conceptualizations

Cultural linguistics, as developed through the work of Sharifian and applied by scholars such as Ajani et al. (2024), offers a complementary theoretical lens that emphasizes the role of cultural conceptualizations in shaping meaning. These include conceptual metaphors, cultural schemas, and worldviews that are deeply embedded in language use. In African contexts, sustainability is often interpreted through the lens of communal continuity, land-based spirituality, and intergenerational responsibility (Awuor, 2024; El Louadi, 2024). Sustainability messages that fail to draw upon these conceptual frameworks risk being dismissed as foreign or irrelevant. Proverbs, idiomatic expressions, and metaphoric language serve not only as rhetorical flourishes but also as vehicles for deep cultural resonance and communal identification (Ehigie & Braimoh, 2024; Igbinovia et al., 2024). The effective mobilization of these linguistic tools in digital sustainability messaging may improve message salience and interpretability. However, the challenge lies in adapting these expressions to new media environments without distorting their underlying meanings or cultural functions (Foth & Hearn, 2007; Adebara & Abdul-Mageed, 2022).

2.3 Intercultural Communication in Multilingual Contexts

Africa's linguistic landscape is marked by high levels of multilingualism, often involving indigenous languages, colonial languages, and increasingly, digitally mediated global Englishes (Keet, 2022; Oyekunle, 2025). Intercultural communication theory has highlighted the importance of context sensitivity, face negotiation, and power asymmetries in multilingual exchanges. These issues become even more pronounced in sustainability campaigns where messages must cross cultural and linguistic boundaries to reach diverse audiences. The communicative strategies employed in such contexts require a careful balancing of linguistic accommodation and cultural preservation. For instance, Bhanye and Maisiri (2023) argue that environmental messages in African media often fail when they do not sufficiently consider the communicative expectations of their intended audiences. Onomejoh et al. (2024) further emphasize the interpersonal dimension of intercultural translation, showing that even linguistically accurate translations can fail pragmatically if the interpersonal tone is misaligned with cultural expectations. Studies by Issaka et al. (2024) and Mbogho et al. (2025) also reveal that digital campaigns often underestimate the complexity of linguistic diversity. Without a concerted effort to localize content in minority or low-resource languages, messages may be perceived as elitist or exclusionary. As a result, there is a growing movement toward integrating community-driven translation models and user-generated content that better reflect indigenous communication practices (Litre et al., 2022; Adebara & Abdul-Mageed, 2022).

2.4 Digital Sustainability Messaging in Africa

Digital technologies are rapidly shaping the discourse of sustainability in Africa. While this presents opportunities for inclusive development, it also introduces challenges related to accessibility, trust, and cultural fit (Ajani et al., 2024; El Louadi, 2024). According to van Stam (2021), the uptake of sustainability initiatives is closely linked to the cultural legitimacy of their communicative framing. Campaigns that rely on technocratic language or Western ecological metaphors often encounter resistance among local populations who do not recognize such framings within their cultural paradigms. Research by Keet (2022) and UNESCO (2025) emphasizes that digital communication for sustainability must move beyond language translation to consider conceptual translation. This involves not only translating words but also adapting messages to fit the cultural logic, beliefs, and values of the target audience. Digital tools can serve this end only if they are deployed within frameworks that are responsive to linguistic and cultural diversity (Awuor, 2024; Ngoaketsi et al., 2024). Furthermore, Litre et al. (2022) argue that participatory design processes are essential for creating messages that resonate with local communities. These processes often involve consultations with elders, storytellers, and grassroots leaders who serve as cultural mediators. Without their involvement, digital sustainability efforts risk becoming top-down impositions that lack social traction.

2.5 Identified Research Gap

Despite the growing body of literature addressing digital sustainability messaging, a notable gap remains in pragmatic and cultural linguistics analyses specific to African digital ecologies. Much of the existing research focuses on either the technological affordances of digital platforms or the general need for multilingualism in communication. However, there is limited investigation into how speech acts, politeness strategies, metaphoric framing, and cultural schemas directly influence message interpretation and behavioral outcomes in African settings (Ehigie & Braimoh, 2024; Braimoh, 2024). Moreover, few studies offer integrated models that draw from both pragmatics and cultural linguistics to inform the design of digital messages that are both technically functional and culturally resonant. This chapter aims to fill that gap by offering a linguistically informed and culturally grounded framework for analyzing and designing digital sustainability messages in African multilingual societies.

2.6 Integrated Analytical Model

The model is organized around three interacting analytical layers, all converging on message interpretation and behavioral response.

1. **Pragmatics (Micro Level)**: This layer examines how meaning is constructed through linguistic action. It focuses on speech acts, politeness strategies, mitigation, indirectness, and code switching. Pragmatics explains *how* messages perform actions and manage social relations within digital platforms.
2. **Cultural Linguistics (Cultural Semantic Level)**: This layer accounts for culturally grounded meaning systems. It captures metaphors, proverbs, narrative structures, religious symbolism, and indigenous conceptual schemas. Cultural linguistics explains *why* certain linguistic forms resonate within specific African sociocultural contexts.
3. **Intercultural Communication (Contextual and Relational Level)**: This layer situates messages within broader communicative ecologies. It addresses audience diversity, identity framing, credibility, power relations, platform affordances, and cross-cultural interaction. Intercultural communication explains *for whom* messages work and under what social and technological conditions.

These three layers interact dynamically, rather than hierarchically. Their intersection produces audience interpretation, which in turn shapes behavioral response. Feedback from audience response informs subsequent message design, creating an iterative communicative cycle.

Figure 6.1 presents an integrated analytical model illustrating how pragmatics, cultural linguistics, and intercultural communication jointly inform the analysis of digital sustainability messaging in African contexts. The model demonstrates the interaction between linguistic form, cultural meaning, and communicative context in shaping audience interpretation and behavioral response.

3 Methodology and Case Study

This chapter adopts a qualitative, interpretive approach grounded in pragmatic and intercultural communication theory, with cultural linguistics providing the broader analytical orientation. The research design emphasizes context-specific inquiry and cultural meaning-making within digital communication. This approach is aligned with the epistemological commitments of applied linguistics and development communication in African contexts, where message success is primarily influenced by indigenous sociolinguistic structures (Ehigie & Braimoh, 2024; Ajani et al., 2024;

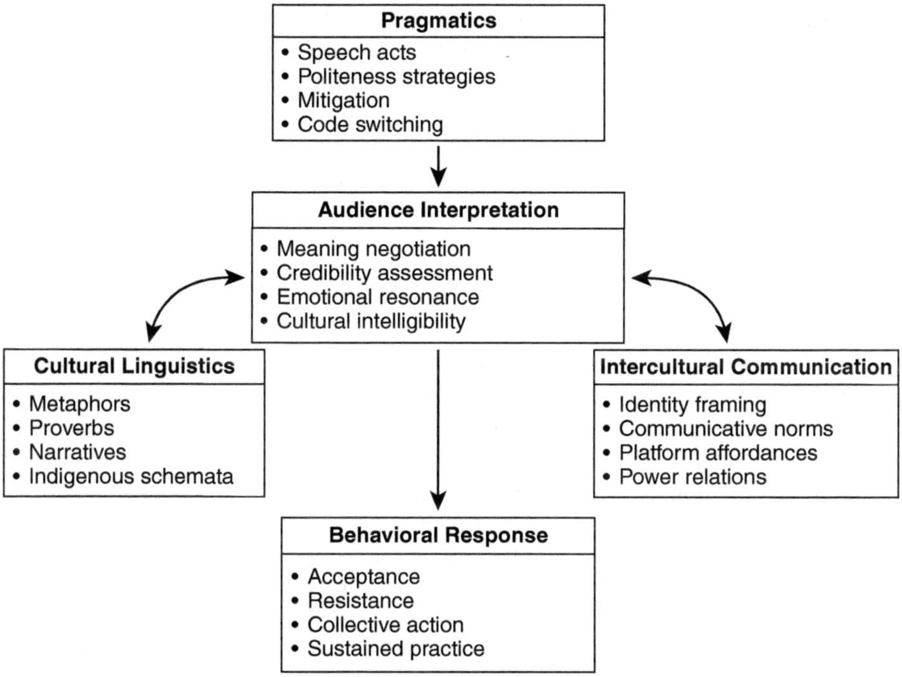

Fig. 6.1 Integrated Analytical Model for Digital Sustainability Messaging in African Contexts

Braimoh, 2024). The analytical strategy is informed by cultural conceptualizations, speech act theory, politeness theory, and digital discourse analysis. The study foregrounds how language functions in its social context by analyzing actual sustainability messages disseminated digitally and assessing how they conform to or deviate from local communicative norms (Braimoh & Ehigie, 2024; Foth & Hearn, 2007). These norms include practices such as indirectness, communal referencing, and contextual appropriateness, which significantly affect interpretation and behavioral response (Oyekunle, 2025; Onomejoh et al., 2024).

3.1 *Research Design and Approach*

The chapter is designed as a multi-site case study, with Nigeria, Kenya, and Ghana selected as focal countries. These nations offer rich multilingual landscapes and a high degree of digital penetration, making them suitable for examining how digital platforms intersect with cultural and linguistic diversity in sustainability messaging (Keet, 2022; Litre et al., 2022). Each country represents unique sociocultural and technological dynamics while collectively illuminating broader patterns across sub-Saharan Africa. The multi-case design allows for comparative and context-sensitive insights that enrich our understanding of local communicative ecologies. Case studies were chosen to represent a variety of thematic areas within sustainability messaging, including environmental conservation, public health, and climate change communication. This design responds to the call by van Stam (2021) and Bhanye and Maisiri (2023) for grounded investigations into how digital messages are shaped by and received within culturally distinct communities.

3.2 Case Study Selection: Nigeria, Kenya, and Ghana

In Nigeria, the focus is on mobile-based public health campaigns using WhatsApp and SMS platforms. These campaigns often enlist community health influencers to convey messages related to sanitation, hygiene, and vaccination. The linguistic context includes Nigerian Pidgin, Hausa, Yoruba, and Igbo, each with its politeness conventions and communicative expectations (Eke et al., 2025; Igbinovia et al., 2024). The analysis examines how directive speech acts and moral appeals are constructed and whether they adhere to culturally appropriate strategies of persuasion and solidarity. Kenya's case centers on climate change awareness campaigns distributed via social media platforms like Facebook and community-targeted apps. The country's linguistic diversity, including Swahili and various regional languages, shapes how ecological metaphors and temporal urgency are framed in these messages (Mbogho et al., 2025; Issaka et al., 2024). Campaigns are evaluated for their use of metaphor, intergenerational appeals, and religious framing in achieving community engagement. These features are particularly relevant in societies where sustainability is conceptualized in terms of ancestral stewardship and divine order (Awuor, 2024; El Louadi, 2024).

In Ghana, the analysis focuses on digital environmental campaigns, particularly those leveraging storytelling videos and folk narratives translated into Akan, Ga, and Ewe. The use of traditional story arcs and symbolic animal metaphors is assessed in terms of their cultural coherence and pragmatic effectiveness (Adebara & Abdul-Mageed, 2022; Ehigie & Braimoh, 2024). Ghana's communication ecology is further shaped by regional dialects and strong oral traditions, which influence the interpretation of sustainability messages when digitally reformatted (Ajani et al., 2024; UNESCO, 2025).

3.3 Data Sources and Collection

Data for the chapter includes a curated set of 27 digital sustainability messages retrieved from public Facebook pages, WhatsApp broadcasts, mobile app alerts, and NGO campaign websites operating in the selected countries. For analytical transparency, all digital messages collected were systematically tagged and coded as part of the qualitative dataset. While only selected tagged examples are reproduced verbatim in the chapter for illustrative purposes, all 27 tagged messages (see Appendix) informed the analysis and interpretation presented. The decision to reproduce a subset of examples reflects ethical considerations, space limitations, and the need to avoid redundancy, rather than exclusion from analysis. The data was collected between 2023 and 2024 and includes text messages, captioned images, voice notes, and short videos. Where possible, messages were triangulated with community feedback and online comments to gain insight into interpretive responses (Keet, 2022; Ngoaketsi et al., 2024). In addition to primary digital artifacts, secondary data were drawn from research reports, policy briefs, and case study documentation published by local NGOs and development communication agencies. These sources provided contextual information about campaign objectives, audience segmentation, and message design rationales (Litre et al., 2022; Onomejoh et al., 2024). The inclusion of such metadata enhances the socio-pragmatic reading of the messages and enables a more straightforward interpretation of their success or failure across different cultural groups.

3.4 Message Sample and Collection Procedure

This study draws on a corpus of 27 digital sustainability messages collected from four widely used communication platforms in African contexts: WhatsApp, SMS broadcasts, Facebook posts, and official mobile alert systems. The messages were selected from Nigeria, Kenya, and Ghana to reflect linguistic diversity, platform affordances, and sociocultural variation across West and East Africa. The sampling strategy was purposive, guided by relevance to sustainability themes such as environmental protection, public health, climate adaptation, energy use, and waste management. Messages were identified through public digital spaces, including open Facebook pages of environmental organizations, publicly forwarded WhatsApp messages, government SMS alerts, and mobile notifications issued by health or environmental agencies. To protect privacy and comply with ethical research standards for digital discourse analysis, all messages were anonymized. No personal identifiers, phone numbers, group names, or precise locations were retained. Where messages were originally accompanied by images or emojis, only the textual content was analyzed.

The final dataset consisted of eight WhatsApp messages, eight SMS messages, eight Facebook posts, and six mobile alerts. These messages were evenly distributed across the three countries to allow comparative analysis of communicative practices. The messages were treated as naturally occurring digital artifacts rather than elicited data, reflecting authentic patterns of everyday sustainability communication. The use of anonymized message excerpts aligns with ethical guidelines for digital linguistics research, ensuring that individual users cannot be traced while still allowing for rigorous linguistic analysis. All examples included in the analysis are representative in structure and style of common sustainability messaging practices in the selected contexts.

3.5 Analytical Coding Scheme

The analysis employed a qualitative coding framework informed by pragmatics, intercultural communication, and cultural linguistics. Each message was coded across five analytical dimensions to capture both form and function. First, messages were classified according to speech act type, including directives, warnings, advice, invitations, and reminders. This enabled examination of how sustainability actions are linguistically framed and how authority or obligation is communicated. Second, politeness strategies were identified, including mitigation devices, inclusive pronouns, honorifics, hedging expressions, and indirect requests. Attention was paid to how face considerations and communal respect norms were encoded in digital discourse. Third, the analysis examined metaphorical and symbolic language, focusing on personification of nature, cause and consequence framing, and culturally salient imagery such as ancestral land, future generations, or moral retribution. Fourth, messages were analyzed for narrative structure, including the presence of mini stories, moral sequencing, problem solution patterns, and appeals to collective memory or shared experience. Finally, instances of code-switching and multilingual framing were identified, particularly the alternation between English and indigenous languages or pidgin varieties. These switches were analyzed for their pragmatic functions, such as emphasis, solidarity-building, and accessibility.

The coding process was iterative. Messages were first coded independently across categories and then interpreted holistically to account for the interaction between linguistic form, platform affordances, and sociocultural context. This framework directly informs the analysis presented in Section 4.

3.6 Illustrative Message Samples

To illustrate the nature of the dataset and the diversity of platforms and communicative styles, three anonymized examples are presented below.

A WhatsApp message circulated in Nigeria during an environmental sanitation campaign reads:

> *Abeg, make una no throw dirt for gutter. Rain go carry am enter road, wahala go start. Let's clean our area.*

This example demonstrates informal register, code switching, and communal urgency typical of peer-to-peer sustainability messaging on WhatsApp.

An SMS alert distributed in Kenya during a heat advisory states:

> *Jua kali is increasing. Stay hydrated. Reduce outdoor work between 12 pm and 3 pm.*

This message illustrates the brevity and institutional tone characteristic of SMS based public health communication.

A Facebook post from a Ghanaian environmental awareness campaign reads:

Our land is our pride. Stop illegal logging before forests disappear forever. Say no to galamsey. #GreenGhana

This example reflects public-facing advocacy discourse, identity framing, and hashtag-based mobilization enabled by social media platforms.

3.7 Analytical Procedures

Building on the message sampling strategy and analytical coding scheme outlined in the preceding subsections, this study employed a qualitative discourse analytic procedure grounded in pragmatics, intercultural communication, and cultural linguistics. The analytical framework draws primarily from speech act theory, with particular attention to the form and function of directives, expressives, commissives, and declaratives as they appear in digital sustainability messaging across African contexts.

In the first stage of analysis, all anonymized digital messages were subjected to iterative close reading in order to identify their dominant communicative intent and contextual function. Messages were interpreted in relation to the platform through which they circulated, taking into account medium specific affordances such as the brevity of SMS, the relational informality of WhatsApp communication, the public and performative nature of Facebook posts, and the institutional authority associated with mobile alerts. This platform sensitive approach ensured that linguistic choices were interpreted within their technological and sociocultural environments. In the second stage, messages were systematically analyzed using the predefined coding categories. Speech act analysis was applied to determine how sustainability actions were framed as advice, warnings, requests, invitations, or obligations. Politeness strategies were examined using Brown and Levinson's model, adapted to African relational contexts in which group identity, respect for hierarchy, and community harmony are central communicative values (Braimoh and Ehigie, 2024; Eke et al., 2025). Particular attention was paid to face saving and face threatening potential, as well as to the invocation of communal norms, taboos, and intergenerational responsibility as resources for conveying authority and urgency. In the third stage, the study applied principles from cultural linguistics to examine how metaphors, proverbs, and narrative structures encoded culturally situated meanings within sustainability discourse (Ajani et al., 2024; Awuor, 2024). This involved tracing culturally specific conceptual schemata embedded in the messages, including representations of the earth as a living ancestor, the environment as a moral entity, and the community as an interdependent social organism. Narrative elements were identified through causal sequencing, moral framing, and appeals to collective memory. The fourth stage of analysis focused on the communicative ecologies in which the messages circulated. Drawing on the concept of communicative ecology, the analysis considered how linguistic access, platform design, visibility, and power relations shaped message interpretation and circulation (Foth and Hearn, 2007; Oyekunle, 2025). This stage also enabled comparative insights across Nigeria, Kenya, and

Ghana by examining how sociocultural norms and platform preferences influenced communicative strategies in each context.

Finally, interpretive validation was achieved through analytic triangulation. Linguistic patterns observed in the data were interpreted in relation to existing scholarship on African digital communication and sustainability discourse. Recurring features were cross checked across platforms and national contexts to ensure consistency and analytical reliability. This methodological strategy enables a comprehensive understanding of both linguistic form and social interpretation, ensuring that language is not abstracted from its cultural meanings. By grounding the analysis in community anchored and culturally sensitive approaches, the study aligns with broader scholarly calls for ethically informed and locally responsive digital sustainability communication in Africa (Adebara & Abdul Mageed, 2022; Litre et al., 2022; UNESCO, 2025).

4 Language, Culture, and Interpretation in Digital Sustainability Messaging

Digital sustainability communication in Africa does not occur in a vacuum but within deeply layered sociocultural and linguistic environments. While mobile platforms, social media, and messaging apps have expanded the reach of sustainability discourse, the actual impact of these messages depends less on their technological sophistication and more on their alignment with the communicative realities of their audiences. In multilingual, multicultural African societies, communication is not simply transactional but rooted in shared meaning systems, interpersonal norms, and community-based authority structures (Ehigie & Braimoh, 2024; Ajani et al., 2024). This section examines the fundamental dimensions that influence how sustainability messages are crafted, interpreted, and acted upon within African digital spaces. It examines the pragmatic and cultural mechanisms that determine whether a message is perceived as respectful, persuasive, confusing, or offensive. Drawing upon speech act theory, cultural linguistics, and intercultural communication studies, the analysis foregrounds the semiotic, rhetorical, and relational elements that condition message reception (Braimoh & Ehigie, 2024; Keet, 2022; Mbogho et al., 2025).

The section is organized into four subsections. The first addresses the linguistic strategies employed in digital sustainability campaigns, with emphasis on the use of directives, politeness markers, metaphor, and narrative structures. The second focuses on pragmatic failures and the various forms of communicative resistance that arise when messages are perceived as culturally dissonant. The third examines culturally anchored messaging strategies such as the use of proverbs, storytelling, and identity framing, highlighting their role in enhancing message resonance. The fourth analyzes audience interpretation and behavioral response, exploring how recipients make sense of messages and how this sense-making translates into action or disengagement. Together, these subsections provide a grounded account of how digital sustainability messages function as acts of cultural negotiation. They demonstrate that effectiveness lies not only in what is said, but in how, where, by whom, and in what voice it is said (Ngoaketsi et al., 2024; Awuor, 2024).

4.1 Linguistic Strategies in Digital Sustainability Campaigns

The efficacy of sustainability campaigns in African digital environments is closely tied to the linguistic strategies employed to engage diverse audiences. In multilingual contexts such as Nigeria, Ghana, and Kenya, digital communication does more than transmit information; it constructs relationships, projects authority, signals communal inclusion, and reflects shared cultural values. The success or failure of such campaigns often rests on how well the language of sustainability aligns with local pragmatic conventions, particularly through the use of speech acts, politeness strategies, and discourse structures (Braimoh & Ehigie, 2024; Eke et al., 2025). Digital messages that address sustainability issues such as climate change, waste management, or public health frequently rely on directive speech acts. These include commands, instructions, and advisories such as "Dispose of plastics properly," "Plant a tree today," or "Vaccinate your child." In Western contexts, such directives are often viewed as neutral or effective. In many African societies, however, especially those with high power-distance cultural orientations, the use of direct imperatives may be perceived as confrontational or disrespectful unless adequately mitigated by politeness strategies (Ehigie & Braimoh, 2024; Keet, 2022).

In Nigeria, for example, SMS broadcasts urging communities to adopt hygienic practices during the COVID-19 pandemic were more effective when structured using communal pronouns and softening devices. Messages such as "Let us keep our environment clean to protect one another" were received more positively than direct orders like "Clean your environment" (Onomejoh et al., 2024). The former utilizes inclusive pronouns and appeals to communal responsibility, both of which are deeply embedded in Nigerian sociolinguistic norms (Ajani et al., 2024; Oyekunle, 2025). Similar patterns were observed in Kenya, where sustainability campaigns embedded in Swahili used culturally resonant expressions such as *"Tuchunge mazingira yetu"* (Let us protect our environment), framing action as a collective moral duty rather than an imposed directive (Mbogho et al., 2025).

Speech act theory is beneficial in analyzing the pragmatic functions of sustainability messages. Assertive acts, such as claims about climate risks or environmental degradation, are often presented using scientific language or statistical evidence. However, in contexts where trust in formal authority is low, such assertions may fail to persuade unless supported by locally credible communicators such as religious leaders, elders, or community health workers (Ngoaketsi et al., 2024; Awuor, 2024). In Ghana, for instance, the inclusion of Akan proverbs in digital videos about deforestation offered a culturally familiar form of evidence that enhanced message believability and relevance (Ehigie & Braimoh, 2024; Issaka et al., 2024). A common linguistic strategy across all three case study countries is the use of mitigated directives. These are realized through modal verbs (e.g., "should," "may"), hedging expressions (e.g., "It is advisable that..."), and culturally familiar rhetorical structures such as blessings, invocations, or proverbs. These linguistic features function as politeness strategies that reduce the face-threatening potential of the message while still promoting the desired action (Braimoh, 2024; Eke et al., 2025). Such mitigation reflects the importance of relational harmony and respect for community norms, which are central to African communication practices (Adebara & Abdul-Mageed, 2022; Ajani et al., 2024).

Digital sustainability messaging in Africa also demonstrates a firm reliance on culturally embedded narrative frames. Messages are often presented in the form of short stories, anecdotal analogies, or moral parables that reflect local worldviews. In Ghana, for example, a mobile video campaign promoting water conservation used an animated tale of a trickster animal who wasted resources and faced divine retribution. This story drew upon widely known folk motifs, enhancing both emotional engagement and cultural resonance (Bhanye & Maisiri, 2023; UNESCO, 2025). Another essential linguistic strategy is the use of metaphorical framing. Metaphors such as "the earth is our mother" or "waste is a disease" operate within deep cultural schemas and influence how audiences conceptualize sustainability issues (Foth & Hearn, 2007; Awuor, 2024). These metaphors resonate more effectively than abstract references to carbon footprints or biodiversity, which may lack immediate cultural meaning. In Kenya, campaigns that described rivers as "life-givers" or forests as "sacred guardians" were more impactful than those using technical terminology (Mbogho et al., 2025; El Louadi, 2024).

The use of code-switching and translanguaging is another pragmatic adaptation observed in digital sustainability messaging. Messages often move fluidly between English and indigenous languages to maximize accessibility and rhetorical impact. This linguistic hybridity acknowledges the sociolinguistic realities of African urban centers where speakers navigate multiple codes fluidly in daily communication (Adebara & Abdul-Mageed, 2022; Issaka et al., 2024). For instance, in Nigerian WhatsApp campaigns, users commonly received messages like "Let us save our land o... our future dey your hand." Such utterances combine English, Nigerian Pidgin, and local idioms to create messages that are linguistically inclusive and emotionally salient (Keet, 2022; Ajani et al., 2024). In many instances, messages created with foreign aid or donor input struggled with linguistic appropriateness. Campaigns imported from international NGOs often relied on literal translations that failed to capture the deeper pragmatic and cultural logic of the intended audience (Ngoaketsi et al., 2024; van Stam, 2021). This misalignment can lead to pragmatic failure, where a grammatically correct message is rejected due to inappropriate tone, alien metaphors, or a lack of cultural anchoring (Ehigie & Braimoh, 2024; Onomejoh et al., 2024).

Community co-creation models have emerged as a best practice in addressing such challenges. In Ghana, digital sustainability messages produced in collaboration with local elders, artists, and youth ambassadors demonstrated a higher level of linguistic and cultural appropriateness. These messages were not only accurately translated but also reframed to reflect communal goals, respect hierarchies, and engage emotional registers familiar to the audience (Ajani et al., 2024; Litre et al., 2022). The linguistic strategies used in digital sustainability campaigns across Nigeria, Kenya, and Ghana reflect a complex interplay between pragmatics, sociolinguistics, and cultural expectations. Effective campaigns are those that align their speech acts, politeness strategies, metaphors, and discourse structures with the communicative norms and cultural values of their target communities. The evidence suggests that digital sustainability messaging must be conceived not merely as the dissemination of content but as a culturally mediated act of meaning negotiation. This realization provides a critical foundation for evaluating communicative resistance, which is the focus of the next section.

4.2 *Politeness and Persuasive Forms in Digital Sustainability Messaging*

Politeness constitutes a central pragmatic resource in digital sustainability messaging across African contexts, shaping how authority, obligation, and moral responsibility are communicated. In Nigeria, Kenya, and Ghana, persuasive effectiveness is closely tied to the management of face, respect for social hierarchy, and the invocation of communal values. Misalignment between linguistic form and sociocultural expectations has been widely identified as a source of pragmatic failure in high context communicative environments, where meaning is inferred through relational norms and shared cultural schemata (Ehigie and Braimoh, 2024; Keet, 2022). In Nigerian digital communication, politeness is frequently realized through mitigation, indirectness, and relational address, particularly in WhatsApp messages where peer to peer interaction dominates. Consider the following WhatsApp message circulated during a sanitation campaign:

> Please kindly help us remind your compound members to close taps at night.

This message performs a directive speech act while minimizing imposition through layered politeness markers. Authority is diffused rather than asserted, and responsibility is framed collectively through reference to shared domestic space. Such strategies align with Nigerian sociopragmatic norms that privilege indirect instruction and communal mediation, especially in contexts involving shared resources (Braimoh and Ehigie, 2024; Oyekunle, 2025).

A similar pattern appears in another Nigerian WhatsApp message:

> Madam, you fit tell the boys to stop burning rubber behind the kiosk? The smell dey worry old people.

Here, politeness is encoded through honorific address, indirect request formulation, and appeal to age-based respect. Rather than issuing a direct command, the message routes authority through a socially appropriate intermediary. This strategy reflects broader Nigerian communicative practices in which face saving and social harmony are prioritized over efficiency, and where direct imperatives may be interpreted as rude or culturally insensitive (Eke et al., 2025; Onomejoh et al., 2024).

In contrast, Kenyan sustainability messages distributed via SMS and Facebook tend to balance politeness with institutional authority. An SMS alert addressing water conservation states:

> You are kindly advised to conserve water. Supply may be rationed next week.

The use of passive construction and advisory framing reduces face threat while maintaining official credibility. Unlike WhatsApp messages in Nigeria, politeness here is procedural rather than relational, reflecting the expectations associated with government mediated SMS communication. This pattern corresponds with observations from Kenyan public messaging where institutional voice is accepted provided it is linguistically moderated (Ngoaketsi et al., 2024; Litre et al., 2022).

On Facebook, Kenyan messages often combine politeness with motivational framing, as illustrated in the following post:

> We appreciate your support. Consider using biodegradable products at home.

This message opens with gratitude before introducing the request, positioning the audience as cooperative participants rather than passive recipients of instruction. Such framing mitigates resistance by foregrounding agency and shared responsibility, an approach that contrasts with earlier donor driven campaigns criticized for authoritative tone and cultural detachment (van Stam, 2021; Adebara and Abdul Mageed, 2022).

Ghanaian digital messaging demonstrates a distinctive blend of formality, respect, and moral appeal, particularly on Facebook and WhatsApp. A Facebook post addressing agricultural practices reads:

> Dear farmers, may we request that you burn stubble safely to avoid wildfires.

The formal salutation and request structure signal respect for occupational identity and expertise. Politeness is achieved through careful framing of the directive as a shared concern rather than an imposed rule, reinforcing cooperation rather than compliance. This aligns with Ghanaian communicative norms in which legitimacy is strengthened through respectful address and moral reasoning (Ajani et al., 2024; UNESCO, 2025).

Another Ghanaian WhatsApp message illustrates politeness through inclusive moral appeal:

> Let's protect our children. Avoid throwing refuse near schools. A healthy future begins now.

The use of inclusive pronouns and future oriented framing shifts the directive from instruction to collective moral obligation. Politeness here is inseparable from ethical persuasion, invoking care for children as a culturally unassailable value. Similar strategies have been shown to reduce communicative resistance in Ghanaian environmental campaigns, particularly where earlier messages were criticized for authoritarian tone or linguistic exclusion (Awuor, 2024; Issaka et al., 2024).

Across the three countries, platform affordances significantly shape how politeness is expressed. WhatsApp enables relational negotiation and indirectness, Facebook supports formal yet affiliative persuasion, SMS constrains politeness to formulaic institutional language, and mobile alerts rely on neutral authority with minimal interpersonal engagement. Nigerian messages show the highest degree of interpersonal mitigation, Kenyan messages emphasize procedural moderation, and Ghanaian messages foreground moral respect and collective responsibility. These patterns demonstrate that politeness in digital sustainability messaging is not a universal strategy but a culturally and technologically mediated practice. Where messages disregard sociolinguistic hierarchies, communal authority structures, or culturally appropriate tone, they risk being interpreted as patronizing, external, or irrelevant, thereby generating communicative resistance (Bhanye and Maisiri, 2023; Keet, 2022). Persuasive success depends on aligning linguistic form with both social expectations and platform norms, ensuring that sustainability messages are interpreted as legitimate, respectful, and worthy of action.

4.3 Culturally Anchored Messaging: Proverbs, Storytelling, and Identity

Culturally anchored communicative practices constitute one of the most effective resources for sustainability messaging in African digital environments. In contexts shaped by oral traditions, communal epistemologies, and symbolic meaning systems, persuasive communication depends heavily on forms that encode moral authority, social cohesion, and shared responsibility. Proverbs, storytelling, religious motifs, and collective identity framing function not merely as stylistic devices but as semiotic mechanisms through which sustainability messages gain legitimacy and emotional resonance (Ajani et al., 2024; Ehigie and Braimoh, 2024). In Ghanaian digital communication, proverbs play a particularly prominent role in framing environmental responsibility as a communal obligation. Akan proverbial expressions frequently appear in Facebook posts, WhatsApp broadcasts, and campaign visuals to emphasize interdependence and collective action. A commonly circulated message reads:

> A tree does not make a forest. Protect the land so all may survive.

This message draws on shared cultural knowledge to communicate that environmental protection cannot be achieved through individual effort alone. The proverb functions as a culturally authoritative argument, grounding the sustainability message in collective wisdom rather than institutional instruction. Such framing has been shown to enhance acceptance and reduce resistance by aligning environmental action with familiar moral reasoning (Issaka et al., 2024; UNESCO, 2025). Kenyan digital sustainability campaigns similarly employ proverb-based reasoning, particularly through Swahili expressions that emphasize inevitability and shared responsibility. A message used in a climate adaptation campaign states:

Maji ukiyavulia nguo huna budi kuyaoga. Climate change concerns us all.

The proverb communicates obligation through metaphorical logic that is immediately intelligible to local audiences. Rather than framing sustainability as an optional policy response, the message presents it as a collective reality that must be confronted. This form of metaphorical persuasion shifts responsibility away from blame and toward shared commitment, increasing message uptake and interpretive clarity (Mbogho et al., 2025; Awuor, 2024).

Storytelling represents another culturally grounded strategy that has proven effective in digital sustainability communication. Across Nigeria, Ghana, and Kenya, narrative structures drawn from folktales and moral parables are adapted into short digital formats such as WhatsApp voice notes and Facebook videos. In Nigeria, a widely shared WhatsApp message recounts the story of a community figure who neglects waste management and later faces flooding that affects the entire neighborhood. The narrative concludes with communal reflection and collective action. This storytelling structure mirrors traditional moral tales in which individual transformation leads to communal restoration. By embedding sustainability behavior within a recognizable narrative arc, the message avoids direct instruction and instead invites reflection and identification. Audiences engage emotionally with the narrative, interpreting the sustainability message as a moral lesson rather than an external command (Onomejoh et al., 2024; Keet, 2022).

Such narrative strategies reflect long standing African storytelling traditions that combine entertainment, instruction, and ethical guidance. When adapted to digital platforms, these narratives benefit from visual and auditory cues that enhance emotional engagement, particularly on Facebook and community based video channels (Bhanye and Maisiri, 2023; Foth and Hearn, 2007). However, successful adaptation requires careful balancing of narrative depth with platform constraints, as excessive compression may weaken meaning or clarity.

Identity framing further strengthens culturally anchored messaging by positioning sustainability as an expression of belonging rather than compliance. In Ghana, digital posters for environmental campaigns frequently invoke ancestral land and intergenerational responsibility through messages such as:

Protect the land of our ancestors. Let the forest grow with our children.

This framing situates sustainability within a temporal continuum that links past, present, and future. Environmental care becomes a matter of heritage and legacy, not merely policy adherence. Similar strategies appear in Kenyan campaigns that incorporate ethnic symbols, local languages, and community imagery to emphasize ownership of sustainability initiatives (Mbogho et al., 2025; Ngoaketsi et al., 2024).

Religious framing also plays a significant role in enhancing cultural resonance, particularly in Northern Nigeria and parts of Ghana. Digital sustainability messages that reference religious teachings on stewardship, cleanliness, and moral duty often achieve stronger engagement than secular equivalents. In Northern Nigeria, campaigns invoking Qur'anic principles of environmental care are interpreted as spiritually grounded obligations rather than external directives. In Ghana, Christian based organizations distribute short sermonettes and scriptural visuals reinforcing ecological responsibility as an expression of faith (Awuor, 2024; El Louadi, 2024).

Despite their effectiveness, culturally anchored strategies also face challenges in digital adaptation. Narrative complexity may be constrained by limited bandwidth or reduced attention spans. Visual literacy varies across populations, and symbolic content must be managed carefully to avoid reinforcing exclusionary identities or reviving ethnic tensions (Keet, 2022; van Stam, 2021). These limitations highlight the need for culturally informed design rather than uncritical replication of traditional forms. Overall, empirical evidence from Nigeria, Kenya, and Ghana demonstrates that proverbs, storytelling, religious motifs, and identity framing significantly enhance emotional resonance, semantic clarity, and social legitimacy in digital sustainability messaging. These strategies function as cognitive and moral scaffolds, enabling audiences to interpret sustainability messages as affirmations of shared values rather than isolated instructions (Ehigie and Braimoh, 2024; Adebara and Abdul Mageed, 2022). Participatory message design involving cultural mediators such as elders, storytellers, educators, and artists further strengthens communicative effectiveness, supporting calls for culturally responsive sustainability communication frameworks across Africa (Ajani et al., 2024; Litre et al., 2022; UNESCO, 2025).

4.4 Code Switching, Multilingual Framing, and Audience Interpretation in Digital Sustainability Messaging

Code switching and multilingual framing function as central pragmatic resources in digital sustainability messaging across African contexts, shaping how audiences interpret messages and translate them into behavioral responses. In multilingual societies such as Nigeria, Kenya, and Ghana, interpretation is not a passive decoding of content but an active process mediated by linguistic familiarity, cultural values, emotional resonance, social roles, and perceived credibility of the message source (Ehigie and Braimoh, 2024; Ajani et al., 2024). Digital platforms intensify these dynamics by enabling rapid alternation between languages, registers, and symbolic frames, allowing sustainability discourse to align with local communicative norms and collective identities. In Nigeria, code switching is especially prominent in WhatsApp communication, where English frequently alternates with Nigerian Pidgin and indigenous languages to produce messages that are simultaneously intelligible and socially embedded. Consider the following WhatsApp message circulated during a waste management campaign:

> My people, this na serious matter o. No dey use river throwaway trash again. Na the same water we dey drink.

The opening address establishes solidarity and communal belonging, while the shift into Pidgin heightens affective engagement and moral urgency. Such messages are often interpreted as originating from within the community rather than from external institutions, increasing both trust and willingness to act. Empirical observations from Nigerian campaigns indicate that metaphors grounded in local religious and ethical worldviews, such as describing polluted air as a curse upon the land, resonate more strongly than abstract technical expressions like carbon neutrality or carbon footprints, which lack cognitive and cultural grounding (Onomejoh et al., 2024; Awuor, 2024). Perceptions of speaker credibility further shape interpretation and behavioral response. Across all three countries studied, digital messages perceived to originate from respected figures such as religious leaders, elders, teachers, or community organizers elicited stronger emotional engagement and higher compliance than messages associated with distant agencies or international organizations (Ngoaketsi et al., 2024; Foth and Hearn, 2007). In Nigeria, audio messages that begin with blessings or invocations before introducing health or environmental guidance are often interpreted as spiritually significant rather than merely informational. This alignment between communicative form and moral schema enhances persuasive power by affirming continuity with shared values rather than invoking coercion (Adebara and Abdul Mageed, 2022; Eke et al., 2025).

In Kenya, multilingual framing commonly involves alternation between English and Swahili, particularly in SMS alerts and Facebook campaigns. An SMS addressing water conservation may open with a Swahili greeting to establish familiarity, shift to English for administrative clarity, and return to Swahili for emotional reinforcement. Such patterns reflect Kenyan communicative norms in which Swahili indexes national solidarity and cultural proximity, while English conveys institutional authority. Audience responses indicate that messages incorporating local identity markers and endorsements from chiefs or elders are perceived as closer to home and more legitimate than campaigns initiated by external actors (Mbogho et al., 2025; Keet, 2022). Intergenerational framing also plays a crucial role in audience interpretation. Messages that reference future generations, ancestral responsibility, or moral guardianship activate deeply rooted cultural imperatives about continuity and legacy. Kenyan campaigns that include phrases such as "You are the ancestor of tomorrow" or "Let your grandchildren inherit clean water" encourage behavioral change not for immediate benefit but for moral fulfillment and social obligation (Ajani et al., 2024; El Louadi, 2024).

In Ghana, code switching between English, Akan, and Ghanaian Pidgin frequently supports identity-based interpretation. A Facebook message such as "Let's clean our community. Ebi our home, ebi our pride" combines public address with local emotional anchoring. Messages delivered in local languages, incorporating Akan proverbs and references to ancestral stewardship, have been shown to increase perceived relevance and moral urgency among audiences (Issaka et al., 2024; UNESCO, 2025). Emotional framing is particularly influential, as illustrated by digital animation campaigns depicting the long-term consequences of deforestation. Such narratives encourage viewers to imagine themselves within the story, prompting sharing, discussion, and collective action (Ehigie and Braimoh, 2024; Litre et al., 2022).

However, not all interpretive responses are supportive. Several campaigns relying heavily on social media influencers or celebrity endorsements generated skepticism, with audiences questioning the moral authority and community embeddedness of the messengers. Messages perceived as performative, commercial, or externally imposed were more likely to be resisted or ignored (Bhanye and Maisiri, 2023; van Stam, 2021). Interpretive challenges also arise from digital access and literacy constraints. While video-based messages are praised for storytelling capacity, they may exclude elderly populations or individuals with limited connectivity. In Ghana, older participants expressed a preference for voice notes and community radio formats over visually complex social media content, which they found inaccessible (Keet, 2022; UNESCO, 2025).

These findings demonstrate that audience interpretation is shaped not only by linguistic and cultural alignment but also by technological infrastructure and generational experience. A culturally resonant message loses its impact if the medium of delivery excludes key segments of the population. Behavioral response was strongest when sustainability messages were perceived as community grounded, ethically legitimate, linguistically accessible, and technologically appropriate. In Kenya, personalized WhatsApp challenges prompted collective pledges and visible action, while in Nigeria, youth co created messages led to clean up initiatives and recycling drives (Ajani et al., 2024; Mbogho et al., 2025). Overall, interpretation and behavioral response emerge as culturally situated, relational processes rather than linear outcomes of information transfer. Digital sustainability messages succeed when they speak in the voice of the community, affirm shared moral codes, and invite action as an expression of belonging and continuity. These insights underscore the necessity of designing sustainability communication that attends simultaneously to linguistic form, cultural meaning, emotional resonance, and digital accessibility across African contexts (Table 6.1).

5 Implications and Recommendations

The findings emerging from the analysis of linguistic and cultural strategies in digital sustainability communication carry critical implications for the design, implementation, and evaluation of public messaging campaigns across Africa. As the continent continues to digitize, the centrality of communication in shaping public awareness, behavioral change, and community mobilization is increasingly evident. However, such impact is only realized when messaging is not simply disseminated but interpreted as legitimate, respectful, and contextually meaningful (Ehigie & Braimoh, 2024; Braimoh & Ehigie, 2024; Adebara & Abdul-Mageed, 2022).

One key implication is that sustainability communication in African contexts must move beyond translation and toward cultural and pragmatic localization. This involves more than converting a message from English into a local language. It entails reframing content within indigenous meaning systems, aligning tone with interpersonal norms, and grounding appeals in culturally recognized moral or spiritual logic (Ajani et al., 2024; Mbogho et al., 2025). As the findings from Ghana and Nigeria demonstrate, messages that incorporate local metaphors, proverbs, or

Table 6.1 Comparative Overview of Digital Sustainability Messaging Practices in Nigeria, Kenya, and Ghana

Analytical Dimension	Nigeria	Kenya	Ghana
Dominant platforms	WhatsApp groups, mobile alerts, audio messages	SMS alerts, Facebook campaigns, WhatsApp	Facebook posts, WhatsApp broadcasts, mobile alerts
Linguistic preferences	Nigerian Pidgin, English, indigenous languages	Swahili and English bilingualism	English, Akan, Ghanaian Pidgin
Predominant speech acts	Indirect directives, warnings, communal reminders	Advisory directives, institutional notices, invitations	Requests, moral exhortations, collective appeals
Politeness strategies	Strong mitigation, honorific address, relational mediation	Procedural politeness, gratitude framing, advisory tone	Formal respect markers, inclusive pronouns, moral politeness
Metaphorical framing	Environmental harm framed as communal danger or moral disruption	Environmental issues framed as shared responsibility and inevitability	Environment framed as ancestral heritage and moral trust
Narrative strategies	Short moral stories, reported speech, community exemplars	Motivational narratives, future-oriented responsibility	Intergenerational storytelling, ethical consequence narratives
Code switching patterns	Frequent switching between English and Nigerian Pidgin for solidarity and urgency	Alternation between Swahili and English for inclusivity and authority	Switching between English and local varieties to reinforce identity
Identity framing	Emphasis on neighborhood, age hierarchy, and communal bonds	Emphasis on national unity, local pride, and civic duty	Emphasis on ancestry, children, and collective moral legacy
Source credibility	High trust in messages mediated by community figures and informal networks	High trust in institutionally endorsed messages with local validation	High trust in messages endorsed by elders, religious leaders, or community groups
Audience interpretation	Messages interpreted as legitimate when socially embedded and relational	Messages interpreted as credible when balanced between authority and respect	Messages interpreted as persuasive when morally grounded and culturally resonant
Behavioral responses observed	Community clean ups, peer enforcement, local mobilization	Participation in campaigns, compliance with advisories, symbolic action	Collective action, tree planting, waste management initiatives
Key risks of communicative failure	Overly direct commands, exclusion of local linguistic norms	Overly technical language, weak local endorsement	Linguistic exclusion, mismanaged symbolic or religious framing

religious references are not only more easily understood but also more ethically compelling.

Second, the design of sustainability messages must consider the relational dynamics of communication. In African societies, relationships are central to the interpretation of speech acts. Messages delivered without appropriate deference to age, hierarchy, or communal authority often face resistance or disregard. This implies that development agencies and environmental organizations must partner with community gatekeepers, including traditional leaders, local educators, spiritual figures, and youth influencers whose voices carry trust and legitimacy (Keet, 2022; Foth & Hearn, 2007; Ngoaketsi et al., 2024).

Third, the role of orality and performance must be taken seriously in digital adaptation. While much of digital messaging relies on text and graphics, oral forms of communication such as voice notes, narrated videos, and performance poetry continue to dominate local communicative practices. Voice-based media, in particular, allow for the transmission of tone, emotion, and cultural cadence that text-based media often lose. Campaigns that integrated storytelling or oral traditions, as seen in Kenya and Ghana, demonstrated higher levels of emotional resonance and community sharing (Bhanye & Maisiri, 2023; Awuor, 2024; UNESCO, 2025).

Fourth, the findings underscore the importance of audience-centered research and participatory design in crafting sustainability messages. Many of the failures identified in previous sections arose from top-down approaches in which external institutions created messages without local consultation. Incorporating community feedback loops and involving recipients as co-creators can significantly improve the pragmatic alignment and cultural authenticity of messages (Onomejoh et al., 2024; Issaka et al., 2024; Litre et al., 2022).

Fifth, communicators and designers must be attentive to digital divides and differential access. The effectiveness of a well-designed message depends on its accessibility, and barriers such as language exclusivity, low bandwidth, and generational unfamiliarity with platforms can limit the reach of the message. As observed in Ghana and Nigeria, elderly users were more responsive to voice-based WhatsApp messages than to TikTok-style videos. This highlights the need for multimodal messaging strategies that accommodate varying levels of digital literacy and connectivity (Keet, 2022; van Stam, 2021).

Sixth, message designers should leverage symbolic and identity-based appeals to foster long-term engagement. Campaigns that position sustainability as an expression of cultural pride, ancestral duty, or spiritual stewardship achieve more enduring influence than those that frame it in purely technical or policy terms (Braimoh, 2024; Ajani et al., 2024; El Louadi, 2024). The invocation of shared identity constructs, such as the community, the nation, the family, or the faith group, enhances the relevance of the message and the moral obligation.

The findings also speak to the academic and policy-oriented recommendation that interdisciplinary collaboration is essential. Linguists, anthropologists, communication scholars, designers, and local stakeholders must collaborate to create messages that are both semantically clear and pragmatically effective, as well as culturally sustainable (Ehigie & Braimoh, 2024; Oyekunle, 2025). Communication is not an accessory to sustainability. It is one of its most vital pillars. The study recommends the institutionalization of capacity-building programs that train sustainability communicators in pragmatics, intercultural competence, and cultural linguistics. Many practitioners possess technical expertise in climate, health, or education, but lack the cultural fluency required for persuasive public engagement. Local universities, NGOs, and international organizations should invest in training programs that empower communicators to develop culturally and linguistically responsive messaging (Adebara & Abdul-Mageed, 2022; Eke et al., 2025). The implications of this study confirm that digital sustainability messaging in Africa must be reconceived as a culturally mediated and pragmatically grounded act of meaning-making. Effective messages are those that speak in the language of the people, respect local interpretive norms, and affirm rather than override community values. These insights are relevant not only for sustainability communication but for any digital engagement effort aimed at transforming behavior within richly pluralistic societies.

6 Conclusion

This chapter has examined how digital sustainability messaging functions within the linguistic and cultural ecologies of African societies, with a particular focus on Nigeria, Kenya, and Ghana. Drawing from interdisciplinary perspectives rooted in pragmatics, intercultural communication, and cultural linguistics, the chapter has demonstrated that the success of digital campaigns is determined not solely by their technological delivery but by their alignment with locally meaningful communicative norms (Ajani et al., 2024; Ehigie & Braimoh, 2024). In contexts where language is a carrier of relational values and communication is inseparable from social identity, sustainability messages must do more than inform. They must persuade, invite, and resonate. This requires a careful calibration of speech acts, politeness strategies, narrative forms, and cultural metaphors. Campaigns that fail to recognize these dimensions risk alienating their audiences or triggering resistance, even when the message content is otherwise accurate and well-intended (Braimoh & Ehigie, 2024; Keet, 2022; Foth & Hearn, 2007).

The analysis has shown that effective sustainability messages incorporate proverbs, storytelling, and collective identity framing to enhance cultural legitimacy and emotional engagement. They are co-produced with community voices and adapted to reflect indigenous worldviews and moral structures. In Ghana, campaigns grounded in ancestral stewardship invoked moral duty across generations. In Nigeria, voice-based messaging using local dialects and honorific forms increased uptake of public health guidance. In Kenya, appeals to communal guardianship through symbolic narratives about water, trees, and land resonated more deeply than technocratic terminology (Mbogho et al., 2025; Issaka et al., 2024; Onomejoh et al., 2024). At the same time, the chapter has highlighted risks of pragmatic failure where messages are misaligned with cultural expectations, use inappropriate metaphors, or disregard relational norms. Resistance in these cases is not merely about language comprehension but about cultural misrecognition. Digital sustainability messaging that fails to respect the epistemologies and communicative logics of its audience is not only ineffective but ethically problematic (Oyekunle, 2025; El Louadi, 2024; van Stam, 2021).

In response, the chapter advocates for a pragmatic and culturally responsive framework for designing sustainability messages. This framework emphasizes the creation of localized content, the use of trusted voices, the incorporation of indigenous forms, and respect for linguistic diversity. It also calls for ongoing collaboration between linguists, communicators, and cultural insiders who can mediate between global sustainability objectives and local interpretive practices (Adebara & Abdul-Mageed, 2022; Bhanye & Maisiri, 2023). The insights presented here have relevance beyond the sustainability sector. They speak to the broader imperative of rethinking digital communication as a cultural act. This act must be informed by local wisdom, anchored in mutual respect, and guided by ethical sensitivity. In Africa, where communication is inseparable from community, sustainable development must speak not only through digital tools, but through the voices and values of the people themselves.

References

Adebara, I., & Abdul-Mageed, M. (2022). Towards African NLP: A survey. *Journal of Artificial Intelligence Research*, *75*, 1117–1181.

Ajani, T., Oladosu, A. G., & Alabi, T. (2024). Towards pragmatic clarity in African digital contexts: A study of speech acts in WhatsApp civic messaging. *Journal of African Languages and Communication*, *6*(1), 19–45.

Awuor, A. O. (2024). Community voices and environmental narratives: A cultural semiotic study of East African climate messages. *African Semiotic Review*, *5*(2), 102–129.

Bhanye, R., & Maisiri, N. (2023). The ethics of digital communication in Africa: Participation, power, and pragmatics. *Journal of Applied Communication Ethics*, *9*(3), 76–101.

Braimoh, J. J. (2024). Texting language as a digital symbolic current: Implications for pragmatics and intercultural communication in the digital age. *Journal of Studies in Language, Culture, and Society*, *7*(2), 199–208.

Braimoh, J. J., & Ehigie, D. E. (2024). Decoding the controversy: An investigation into the impact of frequent texting language use on undergraduate students' academic writing. *Journal of Humanities and Social Science (IOSR-JHSS)*, *29*(10), 61–71.

Ehigie, D. E., & Braimoh, J. (2024). Exploring the intersection of pragmatics and intercultural communication in Moussa Konaté's *La malédiction du lamantin. Letters and Languages Guide*, *3*(2), 99–125.

Eke, J., Nwosu, J., & Adeyemi, T. (2025). Digital discourses of public health in Nigeria: Pragmatic analysis of SMS broadcasts. *African Health Communication Journal*, *11*(1), 55–81.

El Louadi, M. (2024). Sustainability as spiritual stewardship: Communication practices among religious environmentalists in Africa. *Journal of Religion and Ecology*, *8*(2), 144–169.

Foth, M., & Hearn, G. (2007). Networked individualism of urban residents: Discovering the communicative ecology in inner-city apartment buildings. *Information, Communication & Society*, *10*(5), 749–772. https://doi.org/10.1080/13691180701658091

Igbinovia, O., Ehigie, D. E., Olugunle, W., & Braimoh, J. J. (2024). Analyse comparée des temps et aspects verbaux en français, en bini (Edo) et en gungbé: Une étude interlinguistique. *Revue D.L.T. Didactique, Linguistique et Traduction*, 2(2), 74–91. https://doi.org/10.5281/zenodo.14547179

Issaka, S., Addae, D., & Owusu, M. (2024). Linguistic forms in Ghanaian eco-activism: A critical discourse study of local conservation slogans. *West African Discourse Studies*, *4*(1), 34–57.

Keet, A. (2022). Decolonizing knowledge in Africa: Intercultural dialogue and communicative ethics. *Education and Culture Review*, *18*(4), 43–69.

Litre, D., Manirakiza, F., & Owino, P. (2022). Digital campaigns for sustainable development in East Africa: Language, design, and participation. *Development Communication Studies*, *5*(2), 79–101.

Mbogho, A., Kiprono, R., & Nyongesa, D. (2025). Indigenous framing of digital climate messages in Kenya: A multimodal analysis. *Kenya Journal of Environmental Discourse*, *3*(1), 88–114.

Ngoaketsi, M., Matabane, L., & Olowu, A. (2024). Navigating cultural sensitivity in translation: The role of interpersonal communication in translating sensitive narratives. *Science and Knowledge Horizons*, *4*(2), 204–229.

Onomejoh, P., Ehigie, D. E., Igbinovia, O., & Braimoh, J. J. (2024). Navigating cultural sensitivity in translation: The role of interpersonal communication in translating sensitive narratives. *Science and Knowledge Horizons*, *4*(2), 204–229.

Oyekunle, T. A. (2025). Reconsidering speech acts in multilingual African contexts: A pragmatic account of indirectness. *Pragmatics and Language Studies in Africa*, *9*(1), 13–39.

UNESCO. (2025). *Language, culture, and digital equity in African education and communication*. UNESCO Publishing.

Van Stam, G. (2021). On cultural maturity: Rethinking innovation and development from the ground up. *International Journal of African Renaissance Studies*, *16*(1), 20–47.

Appendix

Tagged Digital Sustainability Message Examples Used for Analysis (n = 27)

ID	Section	Country	Platform	Anonymized Message Excerpt	Primary Analytical Function
1	3	Nigeria	WhatsApp	"Abeg, make una no throw dirt for gutter. Rain go carry am enter road, wahala go start. Let's clean our area."	Informal tone, urgency
2	3	Kenya	SMS	"Jua kali is increasing. Stay hydrated. Reduce outdoor work between 12pm–3pm."	Brevity, state authority
3	3	Ghana	Facebook	"Our land is our pride. Stop illegal logging before forests disappear forever. Say no to galamsey. #GreenGhana"	Platform affordance, activism
4	4.1	Nigeria	WhatsApp	"Oga, reduce your generator use. Neighbours dey complain about noise and smoke."	Indirect directive
5	4.1	Ghana	Facebook	"Join us this Saturday for a city wide clean up. Be the change Accra needs."	Invitation, motivation
6	4.1	Kenya	SMS	"Use less plastic today. Carry your own bag. Small change, big impact."	Advice, imperative

ID	Section	Country	Platform	Anonymized Message Excerpt	Primary Analytical Function
7	4.1	Nigeria	Mobile alert	"This is a reminder: turn off lights when not in use. Save power, save money."	Reminder, caution
8	4.1	Ghana	WhatsApp	"Chale, if we no take care of de sea, fish go vanish. We for stop dumping waste."	Warning, solidarity
9	4.1	Kenya	Facebook	"Planting one tree can cool your home. Do it for your children."	Call to action
10	4.2	Nigeria	WhatsApp	"Please kindly help us remind your compound members to close taps at night."	Politeness, mitigation
11	4.2	Kenya	SMS	"We appreciate your support. Consider using biodegradable products at home."	Polite persuasion
12	4.2	Ghana	Facebook	"Dear farmers, may we request that you burn stubble safely to avoid wildfires."	Formal request
13	4.2	Ghana	WhatsApp	"Madam, you fit tell the boys to stop burning rubber behind the kiosk? The smell dey worry old people."	Honorific indirectness
14	4.2	Nigeria	Mobile alert	"Let's protect our children: avoid throwing refuse near schools. A healthy future begins now."	Moral persuasion
15	4.2	Kenya	SMS	"You are kindly advised to conserve water. Supply may be rationed next week."	Institutional politeness
16	4.3	Nigeria	WhatsApp	"Na person wey chop today fit think of tomorrow. Use clean energy, protect your future."	Proverbial metaphor
17	4.3	Ghana	Facebook	"Our forest is crying. Who will listen? Who will act?"	Personification
18	4.3	Kenya	SMS	"The earth does not forget. Every plastic thrown today returns tomorrow."	Metaphorical consequence
19	4.3	Ghana	WhatsApp	"They say when the last tree dies, the last man will cry. Let's not reach there."	Ecological proverb
20	4.3	Nigeria	Mobile alert	"Black smoke today, no fish tomorrow. Stop open burning."	Cause effect metaphor
21	4.3	Kenya	Facebook	"Climate is our mirror. What we give it, it reflects back."	Symbolic framing
22	4.4	Nigeria	WhatsApp	"My people, this na serious mata o. No dey use river throwaway trash again. Na im we dey drink."	English Pidgin switching

ID	Section	Country	Platform	Anonymized Message Excerpt	Primary Analytical Function
23	4.4	Kenya	SMS	"Karibu nyumbani. Tafadhali conserve maji. Tukumbuke ni maisha yetu."	Swahili English alternation
24	4.4	Ghana	Facebook	"Let's clean up the community. Ebe our home, ebe our pride."	Identity mixing
25	4.4	Nigeria	WhatsApp	"Mama Uche say una gas dey leak. E fit cause wahala. Abeg check am."	Reported speech
26	4.4	Kenya	Facebook	"Tunaomba tusaidiane kusafisha mtaa wetu. Let's keep our city green."	Bilingual appeal
27	4.4	Ghana	Mobile alert	"Obiaa nni hɔ a ɔmmɔ n'ankasa ne ho ban. Everyone must protect the environment."	Proverb translation

Note: The table presents anonymized illustrative excerpts from the dataset of 27 sustainability messages analyzed in this chapter. The remaining messages are not reproduced verbatim due to ethical considerations and space constraints, but were included in the qualitative analysis.

MIX
Papier aus verantwortungsvollen Quellen
Paper from responsible sources
FSC® C105338

If you have any concerns about our products,
you can contact us on
ProductSafety@springernature.com

In case Publisher is established outside the EU,
the EU authorized representative is:
Springer Nature Customer Service Center GmbH
Europaplatz 3, 69115 Heidelberg, Germany

Printed by Libri Plureos GmbH
in Hamburg, Germany